THE PURSUIT OF
ECOTOPIA

THE PURSUIT OF ECOTOPIA

Lessons from Indigenous and
Traditional Societies for the
Human Ecology of Our Modern World

E. N. Anderson

 PRAEGER

AN IMPRINT OF ABC-CLIO, LLC
Santa Barbara, California • Denver, Colorado • Oxford, England

Library of Congress Cataloging-in-Publication Data

Anderson, Eugene N. (Eugene Newton), 1941–
  The pursuit of ectopia: lessons from indigenous and traditional societies for the human ecology of our modern world / E. N. Anderson.
      p. cm.
  Includes bibliographical references and index.
  ISBN 978-0-313-38130-0 (alk. paper) — ISBN 978-0-313-38131-7 (e-book)
1. Human ecology. 2. Political ecology. 3. Applied anthropology. 4. Nature—Effect of human beings on. 5. Indigenous peoples—Ecology. I. Title.
  GF41.A473 2010
  304.2—dc22   2009050184

14 13 12 11 10   1 2 3 4 5

ISBN 978-0-313-38130-0
EISBN 978-0-313-38131-7

This book is also available on the World Wide Web as an eBook.
Visit www.abc-clio.com for details.

Praeger
An Imprint of ABC-CLIO, LLC

ABC-CLIO, LLC
130 Cremona Drive, P.O. Box 1911
Santa Barbara, California 93116-1911

This book is printed on acid-free paper ∞
Manufactured in the United States of America

To the memory of
Patricia Gearheart
James Clay Young
Radek Cecil

THEY are all gone into the world of light!
And I alone sit ling'ring here;
Their very memory is fair and bright,
And my sad thoughts doth clear.

It glows and glitters in my cloudy breast,
Like stars upon some gloomy grove,
Or those faint beams in which this hill is dress'd,
After the sun's remove.

Vaughan, Henry. *The Poems of Henry Vaughan,
Silurist,* Vol I. (E. K. Chambers, ed.) London:
Lawrence & Bullen Ltd., 1896: 182.

Contents

We do not find meaning lying in things
nor do we put it into things, but between
us and things it can happen.
Martin Buber (*Between Man and Man,* p. 36)

Preface

Fishing for yellowtail, you need live mackerel for bait. So, at the mouth of Ensenada Harbor, we stopped to buy some. We were a half mile off the Baja California coast; fog hid its soft brown hills.

An ancient, battered boat lay moored there. Tied to it was a well-smack: a wooden barge with the bottom knocked out and replaced by netting. In this huge floating cage were many small mackerel, swimming slowly. A few floated dead on the surface.

We warped in and tied up to the well-smack. Two dogs—black with white points, evidently littermates—ran up to us, seeking affection. They were in magnificent condition, superbly cared for and obviously well loved.

It became clear that the bait man lived permanently in the ancient fishing boat, with the dogs as companions. Now he came forth, a tall, thin man of perhaps 35 or 40, with a timeless, patriarchal appearance. He had a craggy face. A huge, unkempt brown beard reached to the center of his chest.

He moved in sharp, tense, nervous gestures. His few words were jerky, incoherent, senseless. His Spanish seemed strange; he could have been an Anglo. Our boat crew talked with him in the way that Mexicans—who can be the most gentle, sympathetic people in the world—use for people who are "different" or "special." With their help, he netted bait for us. Then he rushed to get a large cup, and begged some coffee from our galley. Our captain paid him for the bait, pushing the money carefully into his pocket instead of his nervous hand.

As we pulled out, the dogs showed us why they were there: With great barking and enthusiasm, they drove away a sea lion and some pelicans that were making a rush for the well-smack and its vulnerable herring.

The bait man disappeared into his cabin. Gulls closed over the boat in a screaming cloud as we pulled away.

This brief encounter sticks in my mind as no book of philosophy has done. The tragically deranged bait man, the Mexican fishermen, the dogs, the gulls, and the fish were tiny and lost between vast sea and vast sky. Yet the brilliance of their lives shone the sun dark.

When we drop the barriers of fear and convention, when we actually confront our fellow travelers on this rolling world, the intensity of the encounter is as much as we poor mortals can bear. When we see with clear sight, the vision blazes in us as endless fire.

This book is a long essay on human interaction, and on human contacts with the nonhuman realm. I have spent my life trying to understand why people use plants and animals in the ways they do. I now want to give back some conclusions about living with this world of interactions.

This book is a personal work, not simple objective scholarship. The last chapters are exploratory—my attempt to come to grips with some moral issues rarely addressed in the environmental literature. There comes a time when a person has to speak—Michel Foucault's "fearless speech" (Foucault 2001). In China I heard a folktale (see Leys 1985 for a slightly different version) of a man carrying a bucket of water to throw on a raging forest fire. Someone tells him, "You can't put out that fire with just a bucket of water." He answers, "Yes, but I've been a forester, and this forest has sheltered and sustained me all my life. There comes a time when you have to do *something*."

I have drawn very heavily on my research experience, involving years of research in Hong Kong, southeast Asia, and Mexico, and short- to mid-length projects in a couple of dozen other countries. Uncited material in this book is largely from my own work and experience.

One reader called this book "very liberal" (he did not mean it as a compliment). Actually, the theme of this book is that local traditions, local cultures, and local small-scale communities deserve our respect and support, and must be the real place of action in saving the world environment. This view of politics cuts across current political-ideological lines. Both American liberalism and conservatism, for instance, appear to be currently committed to large-scale government and top-down management. Therefore, I draw with equal interest and equal skepticism on radical, liberal, conservative, and libertarian ideas.

Full consideration of politics and ethics is necessary for a book like this, but they cannot be accommodated within any reasonable limits of a publishable volume. I have thus followed the examples of the science journals, and placed supplementary online material on my Web site, www.krazykioti.com; look for *Politics* and *Ethics* there.

Of the countless people I should thank for this work, I will acknowledge by name, here, only my old friend and collaborator Paul Buell, who kindly edited parts of the manuscript, and my wife, Barbara. Dozens of colleagues and hundreds of friends in the 60 countries I have visited have contributed to this work.

The dedication of a book is usually a small and unexplained thing. In the present case, the dedication is all-important.

Nothing, except deaths in his own family, saddens a teacher more than the deaths of his students. Their loss is particularly bitter when they are among the very few students who are not only brilliant and creative but who are also passionately devoted to doing their very best for the world. The three students to whom this book is dedicated died tragically at the beginnings of their lives. They had everything to offer, and a sadly short time to offer it. I can only continue their work: the attempt to understand human ways in order to reduce human suffering. If I cannot do it with their genius or their warm and open spirit, I can at least keep their memory green. I can do no more.

1

People in Nature

Heraclitus says that Nature is a child at play; it gives meaning, but in the manner of a child who is playing, and this meaning is never total.
(Merleau-Ponty 2003:84)

FACING A MESSY WORLD

The argument of this book is simple: Our environmental crisis is not the result of simple overconsumption or irresponsibility. It is the result of lack of solidarity. We have failed to come together to get the job done.

This failure has allowed the primary-production interests to maximize their profits by externalizing the costs of production and by getting more and more subsidies from governments—or by taking over whole governments.

They have then, often, maintained disunion by encouraging demagogic politicians who follow the classic Roman formula of *divide et impera,* "divide and rule." Racism, religious bigotry, gender bigotry, and nationalism have been exploited to prevent people from uniting in communities that could actually require the responsible environmental management we all want and need.

I will establish this case in further chapters. The present chapter sets the scene—describing the environmental crisis—and then reflects briefly on the issue of solidarity.

The title of this chapter contains a double meaning. I am concerned with people in the natural environment. I am also concerned with human nature. Yet, is this a double meaning? Can we conceive of people as detached entities, unrelated to the rest of life? Clearly not. Not only do we depend on other animals and plants for survival and subsistence, we also define ourselves in relation to the rest of animate creation.

Some say we are not fully human when we live in wholly artificial environments. Certainly, the postmodern world of "virtual reality"—empty images that mock the truth—is not friendly to the human spirit. If we lose all touch with nature, we will perhaps lose all touch with ourselves, and die of sheer isolation.

I doubt that this will happen. Even in the cities, trees and dogs remind us that there are other lives on earth. So do smaller animals. Wallace Stevens's mice ask a rhetorical question: "Whoever founded a state that was free, in the dead of winter, from mice?" (Stevens, "Dance of the Macabre Mice," 1952:119).

Truly "the mice are nibbling," and the rats too—the literal and the figurative ones—even among the dream-mice of Disneyland.

"Nature," then, is universal, as cherished love and as feared pest or foe. We do not see it neutrally or objectively. We define it and construct it in many ways. We cannot easily separate emotion from reason, or rational beliefs from irrational ones, when we contemplate food or landscapes or the Endangered Species Act. The sight of a fellow animal, even cooked on a plate or simulated on television, awakens the deepest layers of thought.

Human nature arose within a vastly wider realm—a world of mountains and deserts, inhabited by millions of living species. Our relationships with them are no game of "Self" and "Other." They are life viewing itself—a small part of animate creation viewing the whole. Each is simultaneously self and other. We come to know nature long before society has imprinted on our young minds the arbitrary and defensive *Self* that becomes the prison of the unenlightened.

We can, of course, speak of humans as distinct from other animals and from plants. We construct our social and cultural worlds. We have our distinct sphere. Yet we still depend on plants and animals, not only for food and shelter, but also for symbol and inspiration, excitement and rest.

Thus, caring for other lives is not some minor part of an economic agenda. It is integral to human experience. It follows naturally and necessarily from the act of being human. Consequently, a book about managing the environment may well talk—explicitly or implicitly—about human nature.

When I first went out to study cultural ecology in the field, almost 50 years ago, I expected to find people using plants and animals for food, clothing, and shelter. I expected to find what Marshall Sahlins (1976) called "practical reason": people rationally maximizing their material self-interest by exploiting resources, and then going about their other business. I found something more complex. Practical reason was indeed alive and well, but I was surprised to find that the people doing the reasoning were not detached about it—not rational in the sense of *carefully calculating* or *unemotional*

(Anderson 1996; see also Taylor 2006). Whether they were Northern California farmers, Northwest Coast Indian fishermen, or Chinese peasants, people were intensely involved emotionally with the plants and animals in their environments, and with the ways of life they pursued.

Conventional wisdom led me to expect that emotions would contrast with rationality. Emotional people were irrational people. I found the opposite: The more rationally and successfully people were interacting with the environment, the more emotional they were about it. I should already have known this. In human couples, true love and intimacy are opposed to raw passion and infatuation precisely because true love focuses on caring concern for the other's welfare. This caring concern leads to *rational* (carefully calculated—but not unemotional) helping behavior. Humans are constructed such that love cannot be separated from doing one's best in and for the world. Recent psychological work indicates that humans are also constructed such that threat and loss breed anger and fear. Responses to environment thus range from love to hate and from caring to sadistic destruction. They can also range from impassioned response to cold indifference. "Rational" responses, in the sense of unemotional responses, are what we do not regularly observe in such cases (Damasio 1994; LeDoux 1996; Milton 2002; Taylor 2006).

I also noted that people confronted with clear choices often did self-damaging things. At first, I confidently expected that environmental problems would be solved as soon as the public had information on the issues. Surely, people would be rational enough to cease overfishing, polluting, and otherwise destroying their own futures through environmental abuse. Surely, people would recognize that a beautiful environment was not only more pleasant, but healthier, safer, and more profitable than a devastated one. However, in many cases, people went in with close to "perfect information," still made stupid choices, and suffered.

Most of my early career was in fisheries development, a line of work calculated to breed cynicism. People seem almost unable to control themselves when there are fish to be taken. However, I found many cases of successful management of fish resources. People have also succeeded, at various times and places, in managing trees, crops, water, and wildlife. These successes point up the problem we have in understanding the widespread failures. If people realize the need to conserve, and often act on that realization, why is the world in such a mess?

Classically, this question was explained as a "tragedy of the commons" (Hardin 1968, following ideas of Downs 1957 and Olson 1965). People couldn't bind themselves. The "social contract" envisioned by Thomas Hobbes (1950 [1651]) couldn't work.

However, this theory has not stood up well over time; people often, even usually, team up to manage resources (Ostrom 1991). Hardin learned this over the years, and he wrote an excellent qualification: "The tragedy of the *unmanaged* commons" (Hardin 1991). Selfishness is not as common as we thought. When it does happen, it is often not wasteful or destructive, but rather the "enlightened self-interest" described by Adam Smith (1910 [1776]): the butcher and baker serve us because it is their best way to make a living, and the fisher saves fish for the future.

I first realized something was wrong with the conventional wisdom when I worked in Malaysia in 1970–1971. Instead of a tragedy of the commons caused by lack of control, I found the opposite. The fishery had been well-regulated when it was run by local communities. It was self-destroying in the 1970s, because the government had pre-empted control and failed to enforce its own laws (Anderson and Anderson 1978). This, in turn, was largely because the fishermen were Chinese, and the Malaysian government was dominated by Malay ("Bumiputra") politicians, who were increasingly anti-Chinese (a move that climaxed in the early Mahathir regime in the 1970s). Group rivalry co-opted public policy, and trumped enlightened self-interest.

My major interest lies in trying to explain why people treat the environment responsibly or irresponsibly (by whatever standards they have). I am also interested in the more narrowly economic question of why people sometimes maximize long-term interest, but at other times act only for short-term gains, and sometimes act in an outright self-destructive manner. It is relevant and important to understand economic calculations of direct material self-interest, but such understandings must be supplemented by deeper psychology.

Early in my career, I wrote a small essay called "The Life and Culture of Ecotopia" (Anderson 1972). I coined the term *ecotopia* (from *utopia*) to cover a range of possibilities for ecologically aware societies. I was full of radical rhetoric at the time, but the essay had some good ideas in spite of that. These good ideas, and the word *ecotopia*, were taken up by Ernest Callenbach, and developed far better than I could ever do (Callenbach 1975). But ecotopia didn't happen. The human race has created great art, great literature, and a monumental environmental mess. Surely, the species that gave us Shakespeare, Delta blues, and Chinese landscape painting can do better at conserving species and planning pollution control. Something is wrong with our politics. Environmentalists are caught up in struggles about the nature of environmental ethics and of the perfect society (Norton 1991).

Environmental and resource politics has become increasingly bitter in recent years. The issues are complex (see Ascher 1999; Barry's *Rethinking Green Politics*, 1999, which provides a critique rather similar to what follows here; Switzer 1994). Disputes begin with the classic fight of unrestricted use versus sensible use (see Echeverria and Eby 1995). At the other end of the scale, there is a debate over what to do now that we know that most of the world's "wilderness" has been extensively modified and managed by humans for tens of thousands of years (Callicott and Nelson 1998).

PROBLEM LISTS

In general, there are three types of problems, caused by three different types of backgrounds.

Pollution, including global warming. Solutions must rest on setting limits that will protect health of humans and of organisms and resources that humans want or need. This, in turn, requires more efficient production. Pollution is waste. Every poison would be useful if it were captured and properly used. The knowledge base here is better, relatively speaking, but we still need to know much more, especially about cheap and efficient processes that would throughput less material.

Authors since Murphy (1967) have noted that pollution is usually a *desirable* resource being wasted in a harmful way. Dust is someone's topsoil. Sulfur in acid smokestack wastes became a multi-million-dollar resource when environmental concerns forced firms to scrub it from their smoke. Logging slash choking salmon streams is wasted wood. Most garbage and trash could be recycled, and indeed is in many communities today. Thus, those affected are doubly hurt: not only do they have to bear the burden of pollution and contamination, they have to face the consequences of resource shortage caused by resource waste.

Overusing or overharvesting of renewable, useful resources such as soil, fresh water, fish, grazing, forests, and game. Forests are disappearing, and the future will be bleak without timber for construction or green cover for cooling and for oxygen. Farmland is eroding away or succumbing to urbanization, whereas population and demand for food are steadily increasing.

A new development is purchase by crowded but affluent nations of rich farmland in poor nations. Libya has bought into Ukraine; the Persian Gulf oil states into Pakistan, India, and Indonesia; Korea into Madagascar; China into half a dozen countries; and several countries into the volatile and genocidal Sudan. The crowded get food security. The poor get needed investment in their agricultural sector—for now (Goering and Rodriguez 2008).

The solution to shortage is to learn how much we can take—or how much we can use the resource before hopelessly damaging it—and to set limits accordingly. The background problem is that people really need the resources now, and they cannot always afford to forego use today in hopes of a brighter tomorrow. Most overuse is foolish and wasteful and could easily be stopped, but much is not. Small-scale cultivators, fishers, and grazers are often placed in situations where they must overuse resources—against their better judgment—to stay alive at all.

Biodiversity preservation. This problem is the most intractable of all, because few realize how serious it is, and because its background is complex (see Chivian and Bernstein 2008). It involves not only wild species, but also domestic varieties; not only biotic species, but also habitats, including parks and farmland as well as wilderness. The solutions must be based on establishing preserves, corridors, careful land use principles, habitat restoration, and the like. Also necessary is much more research and education, given that we know little about most species, and we teach even less. Unlike pollution and resource conservation, biodiversity protection is not a simple problem of wasteful overuse. It involves wants, needs, tastes, loves, inclinations, and preferences. It also involves genuine ignorance of payoffs and alternatives. People must actively want diversity, and they must learn what is out there. After all, fishing laws in the United States and other countries are comprehensive, and traditional societies have their own strict codes (Anderson 1996). By ironic contrast, there are as many ducks now as when I was young—because duck hunters have had the good sense to save the ducks' habitats and agree to hunt sustainably.

Preservation of biodiversity should, in concept, be easy: Stop unnecessary destruction of life, and create large reserves. But, even in the United States, progress in this direction is minimal. Amphibians are rapidly disappearing because of chytrid fungus (spread by humans), climate change, and habitat loss. Of around 6000 species, 1896 are endangered as of this writing, and over 100 are extinct (Marris 2008). The entire class Amphibia, which has been around for 350 or 400 million years, may well be gone in a century.

Traditional regimes managed to save all the biodiversity we have today. Sometimes they did it deliberately, managing for maximal diversity of life-forms and habitats, as the traditional South Chinese and Maya did. (This is not to say they were flawless, just that they managed for multiple use, and thus for diversity.) Others did not have enough impact on the land to be a major risk factor. Still others were not such good diversity-preservers, losing much, but not destroying all.

At present, we are losing species at a rate that cannot even be estimated accurately. The Zoological Society of London estimates that populations of wild animals (including aquatic forms) dropped 25–30 percent since 1970, with countless species exterminated (BBC News Online, May 16, 2008). Taxonomy and field biology have been so neglected in recent decades that we have no way of even roughly estimating how bad the problem is. Estimates of the number of insects in the world differ by an order of magnitude, and estimates for numbers of species of nematodes or spiders are not much more than blind guesses.

How do we stop a slow, insidious decline when we have reason to believe it will lead to catastrophe (Chivian and Bernstein 2008)? We certainly cannot make the case that the extinction of a fly or a bug will be significant, but we know from countless bitter case studies that the extinction of enough "insignificant" species will bring the whole system crashing down. Paul Ehrlich has made a famous analogy: Exterminating species is like popping rivets on an airplane. Planes are overengineered so that they can stand the loss of a lot of rivets. But finally there comes the last one—a rivet no different, no bigger than the rest. When it is gone, there just aren't enough left to hold the plane together, and it disintegrates in the air (Ehrlich and Ehrlich 1972, 1996).

Countless cancer, infection, and diabetes cures are known to, or alleged by, indigenous peoples, but almost none of these has been tested well. We have hardly even begun to look at marine animals, let alone molds, bacteria, or other microorganisms.

One biodiversity crisis involves domestic species. The vast majority of world agriculture is devoted to producing only ten crops and four animals. After erosion, urbanization, pollution, and soil exhaustion have taken their toll, we will not be able to live on high-yield grains and potatoes and pigs anymore; we will have to be creative (Pollan 2006). There are tens of thousands of useful species that we know, but barely use.

We depend on peasant, semiwild, and wild varieties of cultivated plants. They provide the "new" genetic strains that we use to breed pest resistance, higher productivity, and pollution tolerance into our crops. Genetic engineering offers some hope for the far future in this regard, but for the foreseeable future we will need to save every variety and subvariety we can get. This is now a widely recognized problem. Indeed, it has been recognized since 1846–1848, when North Europe became so dependent on one potato variety (the "lumper") that a strain of potato blight (*Phytophthora infestans*) specially adapted to lumpers led to millions of deaths and millions of emigrants from Ireland, Germany, Poland, and elsewhere. The "Irish"—actually European—Potato Famine had more effect on history than most wars, let alone most kings (Salaman 1949). Local strains of potato in the Andes and Chile are now carefully sought and conserved to keep us one jump ahead of the potato blight.

In many other cases, however, we are allowing stocks to erode (Collins and Qualset 1998; Pollan 2006). The United States has seen perhaps 7000 varieties of apples grown within its borders. About 5000 of these are now lost (Ripe 1994), though heirloom varieties are now being rediscovered in old orchards. Fortunately, centers and organizations devoted to conserving fruit tree varieties are now working hard to find these. Palomar State Park, California, conserves an old settler's orchard—abandoned probably in the Depression—in which no two trees are alike, and none of the fruit bears much resemblance to modern apples. And a new center on Spitsbergen has banked hundreds of thousands of genetic varieties in the deepfreeze, in a tunnel too deep to be vulnerable to global warming.

For overall ecosystem health and specific benefits, we depend on the species we are losing. These points have been made at great length by authors such as Yvonne Baskin (1997), Gretchen Daily (1997), Paul and Anne Ehrlich (1996), and Clem Tisdell (1999), and they need not be further belabored here. These authors have tried to do the near-impossible: to set a value on biodiversity.

A very few species have rocketed from obscurity to world importance. These include such items as rosy periwinkle (*Catharanthus roseus*), source of anticancer drugs, and the molds that are the sources of penicillin and other antibiotics. The untapped potential is known to be enormous; a huge review (Chivian and Bernstein 2008) discloses countless possibilities, some lost, and some barely explored but acutely endangered. My own research has turned up many valuable medicinal plants not now utilized, but problems with patenting have prevented me from doing anything with them.

This has led several economists to calculate the value of the average plant, the average hectare of rainforest, and so on (Southgate 1998:85–94). Suffice it to say that the figures range by several orders of magnitude. Going with the higher estimates, based on past experience with periwinkle and the molds, we come to very high values: as much as $1.63 million per plant species (Farnsworth and Soejarto 1985; note that would be close to $3 m in today's dollars). But sober estimates of the value of future plant finds run as low as an average of $67 (Southgate 1998:89). This latter figure is too low—it does not build in estimates for the effects of world plant depletion, evolved immunity by germs to existing antibiotics, or anything else relevant—but it does show the range of figures that sober minds can support.

There are three basic strategies for saving biodiversity: save, manage, or restore.

It is reasonable to begin with the Endangered Species Act. This epochal and vital legislation has its strengths and weaknesses.

The law was originally designed to save the bald eagle, the national symbol of the United States. Conservationists of the time (the early 1970s) were also concerned about other noble birds of prey—especially the California condor, always rare, and the peregrine falcon, endangered by pesticides.

Bald eagles need a great deal of water and fish, but not much else except for nesting and roosting trees. The logical thing to do was to protect them as individuals—to ban all shooting, sale, and possession of the birds or parts of them—and then to protect their most critical habitat, such as nest trees. They had the good taste to live largely in wilderness on public land.

Unfortunately, they are also a very atypical endangered species. They range very widely over North America. They are highly adaptable—they will live anywhere, including the middle of a city, as long as water and fish are around. They tolerate any kind of habitat, as long as it has a nest tree—and a rock will do if trees do not afford themselves. Above all, Bald Eagles are charismatic. They are the United States' national symbol.

The vast majority of endangered species are very different. They are animals and plants that were always rare and local. They depend on a relatively undisturbed and pristine habitat. Many of them live on private land, often valuable for development. And most modern Americans think of them as obscure, scruffy, and uninteresting.

This makes saving them difficult, for obvious reasons. In particular, landowners do not take kindly to discovering that property from which they wanted to make a fortune, or at least a decent living, is now off-limits because of a worm or a fly.

Worldwide, the reality is much more disturbing: Hundreds of millions of very poor rural people, for whom any up-front costs may be fatal, come into head-to-head competition with millions of species, most of which are not even studied or described.

There has to be absolutely "zero tolerance" for letting species die out. The danger in allowing obviously useless, even pernicious, ones to go is that the "slippery slope" argument is real and frightening here—in a way it rarely is in politics. Once we start playing God, deciding which species shall live and which shall die, we will find it hard to stop exterminating "just one more." Even the smallpox virus will be needed, to make inoculation material should smallpox arise again by mutation, or should it be unleashed in biological warfare by some nation that has sequestered a secret supply of virus.

However, we cannot place the entire burden of saving these organisms on the backs of the landowners. One of the major needs of the world today is developing proper compensation.

SOME SUGGESTIONS

John Holdren, in his presidential address to the American Association for the Advancement of Science in 2007 (published as Holdren 2008; he is now science advisor in the Obama administration), linked these environmental problems to concerns about poverty, disease, violence, oppression of human rights, waste of human potential, and maldistribution of consumption. Energy, in this age of global warming, is a particular concern. He proposed many ways in which science and technology can address these problems, and recommended not only these but a much broader social focus on science education.

Norman Myers (1998, 1999) makes four key points (Myers 1999:36–38; see also Myers 1998):

1. *"Defuse the population bomb."* This should be so obvious as to deserve no comment. However, claims that we have no problem, or that population will automatically level off around 2050, have been widely maintained without any evidence. The facts are otherwise: Extreme crowding and economic problems from high population worldwide; a world where the poor simply cannot live like the rich, because there are not enough resources; and rapid population growth continuing, with no sign of stopping, in most countries.

   Population increase has come to a halt in many developed countries and is slowing in many poor nations. The reasons are well known: Availability of full contraceptive technology (ideally, at nominal cost); social welfare, so that people see alternatives to having many children for support in old age; and education for girls. The dramatic correlation between girls' schooling and declines in fertility rates has been particularly thought-provoking, and particularly important worldwide. We can cause birth rates to decline, and (ultimately) population growth rates to decline too, without recourse to the horrific methods used in China and, for a time, in India. Yet, with all our knowledge, population continues to grow rapidly in much of the world. Religious bodies continue their adamant opposition to birth control. Many traditional societies, especially in Africa and west Asia, oppose girls' education. All this has stopped the progress of the demographic transition; even the United States is increasing its population at a disturbing rate.

   Stopping population growth was a recognized need in the 1960s (Ehrlich 1968), but it has disappeared from the world radar. Right-wing groups oppose it for various reasons, including a felt need to increase their followers' population in comparison to others.

2. *"Reduce wasteful consumption."* This too is obvious enough. Waste and manufactured "wants" characterize much of the world. The ruin of forests, water, and soil that one sees in Mexico and other Third World countries shows that wasteful consumption is not necessarily related to affluence. In these countries, it leads directly to impoverishment. The Soviet Union, too, devastated its environment without producing affluence.

3. *"Relieve Third World poverty."* The Third World poor are the people who, out of sheer desperation, torch rain forests and wreak all kinds of other

habitat destruction (p. 37; actually the poor do rather little damage, but reducing poverty is necessary for other reasons, as we shall see).

4. *"Get rid of perverse subsidies."* The average American taxpayer shells out $2,500 a year to fund government support for fossil fuels, road transportation, agriculture, water and fisheries. Then the same taxpayer turns around and spends another $2,500 to fix the environmental problems those payments produce and to pay higher prices for consumer products. . . . Worldwide, the total of such rat-hole subsidies amounts to a whopping $1.5 trillion per year. . . . Lobbyists in Washington spend $100 million a month to press their causes, with these subsidies at the top of their lists" (p. 38; the figures are now very much higher).

Myers points out that "during the past ten years, 55 million Americans have given up smoking. That has been a social earthquake, virtually overnight" (p. 38; and now, ten years later, the figure is almost doubled again).

SUSTAINABILITY

Despite disingenuous arguments to the contrary, the basic meaning of this term is quite clear. "Sustainability" is a simple matter when we talk about fish, or deer, or pine trees; we know that the idea is to take few enough of them to permit natural recruitment to replace the loss. Determining the actual number or amount can only be done by trial and error. Models help but are not adequate in themselves. The laws setting seasons, bag limits, and quotas have to be designed according to models but corrected by experience. Moreover, there has to be a provision for complete shut-down if there is a sudden, unexpected catastrophe—a massive epidemic disease, a forest fire. The stock has to recover. This method is maddeningly ad hoc and sloppy to an ecologist or economist, but it has protected game, fish, and forests in the developed world wherever it has been used. It has protected many resources in traditional societies for thousands of years.

Sustainable farming is more difficult to institute, but the idea is similar. The farming is designed to prevent soil erosion, deforestation, desertification, or other problems that would feed back on the wider rural system and harm it. This begs the question of what the "wider rural system" is—just the farms, or the whole scene? So there is room for negotiation. However, this is debate about the actual goal of the people on the ground, not about the nature of sustainability.

Sustainability gets problematic when we turn to such concepts as "sustainable development." Does that mean that the impetus or rate of development can be sustained? Does it mean that development should not impact sustainably croppable resources? Does it mean we can't use nonrenewable resources like groundwater and oil? Does it mean we stick to doing things that we can do indefinitely? Obviously, all these definitions imply different things.

If we mean that development should not lead to net loss of sustainably manageable, renewable resources, then we are on safe ground. This is quite possible, and indeed is occurring in many areas. In Korea, development has gone along with a huge reforestation initiative that has renewed Korea's forests and wildlife (once almost totally lost). In the United States, development in the late 20th century was dramatic, yet forests increased, ducks rebounded after a crash, and deer became a pest in many areas.

If, however, sustainable development means that nothing unsustainable is to be done, the concept becomes truly absurd. Any development is going to use resources. Some of them will be nonrenewable. There is, however, no obvious benefit in leaving deep-buried resources in the earth, unless we know that future generations will value them more than we do. I doubt they will; we certainly don't know. Development, in general, normally proceeds through the process of doing something unsustainable for a while, getting the benefits of that, and reinvesting them in something better. This may be observed in the production-bootstrapping that England and the United States invented and Hong Kong perfected: Every production process is superseded, but not before the producers have realized profits and invested them in something that adds more value to the product. The question is not one of development without *any* costs, but one of development with *bearable* costs. In particular, nothing truly irreplaceable and unique should be allowed to disappear. This refers to human languages and important cultural forms as well as biological species.

CASES: SONGBIRDS AND RURAL PEOPLE

Many worrisome problems are poorly known. Populations of North American, European, and Asian migrant songbirds are declining at a rate of one to four percent per year. Long-distance migrant shorebirds are declining even more rapidly. Sea birds decline due to overfishing and pollution.

The canaries, the literal canaries, are dying all around us.[1] Many of the toughest, most resilient, and most human-tolerant birds are in catastrophic decline. From bobwhite quail, loggerhead shrikes, meadowlarks, and red-eyed vireos in America (Terborgh 1989) to the incomparable nightingale in Europe, the once-common dooryard birds are dying out (see again Chivian and Bernstein 2008). If even these are in trouble, the world situation is desperate indeed.

I am sad to see the loss of the vast flocks of migrants. My children and their children will never see what I saw in my youth: trees and bushes filled with brilliant, flashing colors, ten or twenty or fifty migrant warblers and tanagers in every tree or bush over a whole forest or mountain. In California, migrant rufous hummingbirds used to fill the lowland groves in spring and the mountain meadows in late summer. One could watch hundreds of them at a time, darting back and forth, shining red in the sun. Now, one is lucky to see even one. Where I used to count thousands of shorebirds and seabirds off Malibu or in Puget Sound, I now find at most a few dozen.

The decline of songbirds parallels and warns us of the decline—far more sad to humans—of the human poor. This has been concealed behind meaningless statistics of "economic growth" that average a few rich urbanites with several millions of desperate small farmers and landless laborers. It is the latter that bear the full brunt of soil erosion, fresh water exhaustion, loss of cheap fish (often their protein staple in the past), deforestation, chemical contamination of the environment, poor health care, unavailability of birth control resources . . . The list goes on.

The one or two billion poorest people on earth (Collier 2007) are largely rural. I lived and worked with such people in the forests of Quintana Roo a generation ago, when some 30 percent of the people in my area of research reported "no income" on the census (Anderson 2005). "No income" meant they were growing tons of maize and other crops, building small but very serviceable houses from local timber and thatch, raising pigs and chickens, collecting wild foods, curing themselves with herbs (their medicines were highly effective, working for me better than drugstore remedies), and enjoying a wonderful life. I found that they would have had to pay about $500 per capita to get as much from the market. In today's dollars, that would be more like $1000. To live a comparable lifestyle in the United States would have been far more costly.

As one got progressively closer to the big cities of Mexico, the population density and extractive effort rose steadily, until this sort of subsistence lifestyle became impossible. Near the cities, the soil was worked out, the firewood gone, the wild herbs overcollected.

In many areas I know and have studied, from Madagascar to Mexico, people who lived good and honorable lives a generation ago through subsistence agriculture are now starving and dying. They had no money before, but they raised their own food, drew on forests and fields for fuel and building materials and herbal medicines, hunted, and fished. Now they have nothing. Infant mortality in such communities remains high, up to 20 percent in Africa. Maternal mortality is so common that, in many areas today, a woman has a 1 in 7 chance of dying in childbirth. (The rate in Scandinavia is about

1 in 30,000.) Whole regions, whole ethnic groups, are endangered. Even where people survive, appalling cruelty and oppression lead to poverty, language and culture loss, demoralization, and violence. Genocide, horrifying as recently as the 1940s, has become a routine aspect of policy, especially in Africa; the international community ignores it. (On all these problems, see Collier 2007; Humphreys et al 2007.)

Particularly disturbing are the recent national meltdowns in which environmental devastation is clearly a factor. The extreme ecological degradation of Ethiopia, Rwanda, Burundi, Liberia, Somalia, Afghanistan, Guatemala, El Salvador, and several other countries, coupled with governmental irresponsibility or venality, clearly had something to do with the civil wars that ravaged them in the 1980s and 1990s (Collier 2007).

The pioneer economy of the 19th and 20th centuries cannot be sustained. By the late 20th century, most sustainable natural resources were showing signs of overuse. Conservation and ecological management became more popular. At the beginning of the 21st century, resources everywhere are stretched to the limit.

Good farmland is getting scarcer, due to soil erosion, salinization, and urban sprawl. If present rates of urban expansion continue, all of California's farmland—the most productive in the world—will be lost by 2050. So far, the world has been able to increase production per acre to keep up with population increase, but this approach is running into problems. The infrastructure, for example, must improve along with the increase in yields; one must have roads to get heavy loads of grain to market, and storage facilities to hold it. Also, the more production increases, the more need there is for inputs such as fertilizer.

Agricultural research is not free. More and more money has to be invested in developing new crops. Only the United States (public and private sectors) has a truly mammoth research agenda. As the United States pushes its own limits, this funding will also fall short. Other developed countries could, and probably will, increase their agricultural research funding, but Third World countries have little money—all the more tragic in that they are usually the homes of the most diverse and interesting new crop strains and of the most challenging local techniques of farming.

During its history, the United States has destroyed much of its forest and wildlife frontier, but much of the profit has gone into railroads, factories, and education systems. The bargain may not have been a good one, and certainly much money has been wasted, but at least some of it was invested in productive enterprises. Many Third World nations have not done that (see, e.g., Collier 2007; Hancock 1991; Southgate 1998). Resource income went into armaments, luxury consumption, and Swiss bank accounts. Now the resources are gone, and there is nothing to show for it but ruined landscapes and ruined people.

CASES: LAKES AS MICROCOSMS

Two small, closed worlds may serve as excellent examples of the problems we face. Some years ago, a well-meaning soul introduced Nile perch to Lake Victoria in East Africa. The Nile perch began to eat the native fish. There were over 400, perhaps over 700, species confined to the lake—how many more, we have no clue, and will never know now (Stiassny and Meyer 1999). So far, the Nile perch has exterminated about 300 of them, but by the time you read this the total may be over 500. Moreover, the cichlids controlled algae, and now that they are gone the algae are multiplying and slowly killing the lake. Nothing can be done now to control the perch. The future is easily predicted from comparable stories elsewhere: The perch will eat itself out of a food supply, turn on its own young, succumb to algae overgrowth, and leave us with few Nile perch—let alone any of the other species.

In Central Asia, the Aral Sea was the fourth largest lake in the world, a slightly salt lake filling a vast basin. In the mid-20th century, the Soviets developed intensive cotton farming along the rivers that fed the Aral. Most of the water went for badly-engineered irrigation, sinking into the sands or evaporating into the hot, bone-dry air. The sea began to dry up. What water did reach the Aral was contaminated with pesticides and inorganic fertilizers. The Soviet system subsidized extensive use of these chemicals, far beyond any real need, far beyond any safe margins. Natural salts, leached from the earth, also entered the water. As of 2008, the sea was 1/10 of its original size (for this and what follows, see Kobori and Glantz 1998; Micklin and Aladin 2008).

As the sea dried, these poisons were concentrated. Soon, its million-dollar fishing industry, landing 40,000 metric tons in 1960 (Micklin and Aladin 2008:66), collapsed. Boats rusted, far from the shrinking body of water.

The wind whipped toxic clouds of dust from the dry seabed. This dust seeped into everything: clothing, houses, food. Soon the children began to die. Poisoned water and poisoned air were the only world they knew. Today, the overall infant mortality rate of the Karakalpak Autonomous Region (south of the Aral Sea) is 10 percent. In a few areas it rises to 50 percent or more. The Karakalpak people are afraid for their very survival. The situation appears to be as bad on the northwest-lying Kazakhstan side of the sea, as well. Health is damaged much farther afield; the toxic dust blows thousands of miles across the Central Asian deserts and steppes and on into the Arctic, poisoning land and humanity as it goes.

The worst-affected region is now in the impoverished, newly independent country of Uzbekistan, which cannot afford to clean up the mess, or even to move rapidly out of intensive cotton agriculture. So the sea continues to shrink, and the land and people continue to die. Eco-catastrophe is here.

Much of this destruction has been done in the name of economic progress. No doubt the Soviet engineers who destroyed the Aral Sea are still proud of their success in irrigating cotton (as the Spanish Inquisitors were proud of their record of burning heretics). Indeed, the world needs more food and fiber. But the costs of badly designed "progress" are real and terrible. Moreover, much so-called progress has merely been the substitution of "modern" agriculture for more productive but less market-oriented agricultures. This happened in the Aral Sea case, where a successful and long-established local agriculture was wiped out to mine the environment for cotton. Modernization of agriculture is worthy of the name "progress" only if it feeds and clothes more people than any alternative system could do.

The Aral Sea continues to decline, but Kazakhstan, which controls the north part, has stabilized the decline there, by dyking off the north from the rest. The north is returning to normal; fortunately, Kazakhstan controls the delta of the Syr, one of the two rivers feeding the sea. (The other, larger one, the Amu, is controlled largely by Uzbekistan, still committed to suicidal irrigation agriculture.) As Micklin and Aladin (2008:71) point out, "Humans can quickly wreck the natural environment, but repairing it is a long, arduous process."

NATURE?

At this point, we need to backtrack to a deeper philosophical issue.

There has been a recent problem with the concept of "nature" that betrays a lack of meeting of minds between biologists and humanistic scholars (Latour 2004).

Biologists tend to see the concepts of "environment" and even "nature" as unproblematic. Nature is what isn't human. People obviously affect and influence it, but it is a separate realm. The problem of delineating where nature stops and humanity starts is a minor one; obviously there is a continuum, and you can cut it where you want. A city is clearly not a natural environment. A modern industrial farm is not really one, either, but it depends on "nature's subsidies" (Daily 1997) in the form of plant varieties, water, soil, and sun. A traditional farm is much "closer to nature." Finally, genuine wild nature is whatever is beyond the farms and ranches. Conservation biologists thus want, or until recently wanted, to save "wild nature" from "human influences" or "human degradation."

This simple view has recently foundered on evidence that almost the entire earth's surface has been profoundly modified by human agency for thousands of years. The oak-hickory forests of eastern North America, once the very definition of "wilderness" for American settlers, now turn out to be created and maintained by fire—to a great (though uncertain) extent, by Native American burning (Delcourt and Delcourt 2004). The Native Americans used fire as a management tool, deliberately burning the forests so as to maximize the number of nut-bearing trees. In fact, if these findings are right, the Eastern North American sylva is better seen as a giant nut orchard than as a wilderness. It appears that people often intervene in such wise as to increase biodiversity, rather than diminishing it. Much has become extinct because of human devastation, and even because of the management strategies. (What got burned out to make those forests?) However, what we have now is what is left after all that, and if we want to save it (and we do) we sometimes have to manage it.

Taking this view to an opposite extreme, humanistic scholars have tended to see "nature" as a cultural construction. Obviously, the English word "nature" is a cultural item, and so is any meaning we attach to it. People use the word in various different ways, and many (if not most) languages lack any equivalent. The Chinese word usually translated as "nature" is *sheng*, or compounds thereof. These terms really refer to one's inborn or inner nature, not (originally) to a whole biotic universe. Maya, the other non-Indo-European language that I know, has no term even remotely close to the English "nature"; the nearest one can get is *ba'alche'* ("things of the forest"). A more relevant concept is *maayab*, "the Maya land," both domestic and wild; the Maya do not see a basic separation.

Other words, such as *wilderness* (possibly from "wild-deer-ness," the place of wild animals) are also cultural artifacts. *Wilderness*, with all the favorable connotations it has rather recently acquired in English, is particularly difficult to translate. In many languages, equivalent words have only bad connotations. *Environment* is another problematic term, variously defined in English and lacking close equivalents in other languages; Spanish had to coin a phrase, *medio ambiente*, to translate it.

To a hard-core humanist, words are things and things aren't. Extreme postmodernism privileges "text" and "discourse" so much that underlying reality is sometimes denied outright. So we find that people deny there is an environmental problem because the words have culturally conditioned meanings. To paraphrase the Buddhists, this is like failing to leave a burning building because *fire* is not precisely defined, or denying that there is any difference between life and death because we can't set a firm boundary and both words have highly culturally constructed meanings. Cultural construction usually makes a word more accurate and useful, not misleading and silly (Latour 2004, 2005).

The findings about long-standing human effects on the environment are (rather paradoxically) grist to the humanist mill. It enables them to say that the environment is a cultural creation in a more real and literal sense. It enables them to dismiss all the rhetoric about saving "virgin nature" and the like. Thus we find Simon Schama, in *Landscape and Memory* (1995), casting doubt on the reality of environmental concerns. Having found that the words shift in meaning, and also that Western civilization has always profoundly modified "nature" and has viewed it in many different ways, he appears to conclude that there is no such thing as "nature," and therefore that all environmentalism is misguided. In fact, he implies very strongly that it is fascistic, because he found some conservationist rhetoric among the Nazi leaders. (Actually, their rhetoric was specious and superficial, and had nothing to do with modern conservation or ecology.) Yet I doubt he would respect the logical corollary of his line of thought: If mere words are all there is, and mere words cannot give us morals or moral truths, then Nazism itself cannot be condemned, and the Holocaust can be disregarded. After all, Nazism and opposition to it were both culturally constructed, and the Holocaust is subject to various interpretations. In a world of words and ideas, we have no reason to disregard those who say the Holocaust never happened or was trivial. In fact, we have no reason to care whether six million people were indeed killed. "Caring" and "morals" are mere words too—and, worse, they label internal states, not tangible items.

"Reality is what refuses to go away when I stop believing in it" (to quote an anonymous tagline floating on the Internet), and if there is a reality, then there are people and there are nonhuman beings. People and other things interact and are changed by the experience.

It does not really matter whether we call the nonhuman entities "nature," or *ba'alche'* (Maya for "things of the forest"), or *medio ambiente*. It does not matter whether we see them as some sort of undifferentiated mass ("nature") or as a cluster of species (as the Maya do). What matters is that we know there are things out there, and that we depend on them, and that we have to manage them better than we are doing if we want to survive.

Yet, the word *nature* may now carry too many unfortunate and misleading implications of an untouched, unaffected place.[2] We are better off with *environment*, a word which can accommodate everything from urban and built environments to the Antarctic wilderness. *Wilderness* does have staying power, so long as we recognize that there are no absolutely pure wildernesses left. We desperately need to preserve wild lands, to maintain not only the biota but also our sanity. If they are not 100 percent wild, so be it; anything with a variety of non-human-created things (however measured) should qualify. Indeed, we need even more desperately to preserve and improve whatever is left of wild spaces in the urban and periurban areas of the world. Drawing on my wanderings in urban Europe, I can testify that Amsterdam's Vondelpark is not pristine, but it has perhaps the highest density of nesting songbirds of anywhere I have visited; and that the Parc Parilly in Lyon, France, planted on what was worked-out farmland (and barely growing when I first saw it), is packed with nightingales, and has been colonized recently by green woodpeckers and other birds of the wild forest.

ALL CAN BE FIXED

The environmental movement began in the 19th century, and has been a major world political force for over 30 years. Today, all polls and surveys in the United States (and other countries, both developed and less-developed), yield the same result: By overwhelming majorities, people want the environment protected, and are willing to pay for it.[3]

Yet Americans do not vote that way. For most of the last 30 years, they have voted for anti-environment candidates and have killed most pro-conservation measures submitted as voter initiatives. To some extent, polls ask questions in a loaded manner, and people answer polls with what they "know they should say," rather than what they really think. Everybody is against sin and for clean water. But in this case, it seems clear that national politics and national behavior really are far behind the public will.

Most people, according to all opinion polls, prioritize immediate concerns such as crime, health care, and military preparedness over longer-term environmental concerns. There is some reason to do so. Immediate problems have to be dealt with immediately. If one is being murdered, it is no consolation to know that there are national parks out there somewhere. Keeping the environment on the agenda is desperately needed but not at all easy to do, even in the United States, let alone in seriously troubled nations such as Russia or Iraq.

Environmental ruin is hardly a new thing in the world. The Roman Empire and at least some ancient Maya states suffered, and may have fallen, because of it. The ancient Near East, old China, and the Polynesian islands, among other areas, faced devastation.[4]

Environmental decline is preventable. We know how to do it. We have key technologies. Compared to a 1950s American car, my tiny new car has about three percent of its ecological footprint—from manufacture (from light materials) through fuel-efficient use to final recycling in the far distant future. My laptop has thousands of times the memory of the first room-filling computer, a machine built within my lifetime. Most of my household wastes are now recycled. In every way, I can (at least potentially) use far less stuff and get far more satisfaction from it.

On a worldwide scale, we could reforest with fruit trees, raise bison instead of cattle on the high plains, fish sustainably, treat contaminated water and use it in drip irrigation, develop solar power, eat locally, and so on and on. We have all the technology we need to save the world.

Soil erosion control, for instance, is no great mystery; it was successfully applied during the 1930s in the United States. Yet it remains beyond the apparent scope of most governments worldwide. The United States itself still allows much unnecessary erosion to occur. Mexico, where I do my current research, has shown even less concern; much of the country is now a lunar landscape. More of it is becoming so every day, as deforestation is followed by up-and-down plowing on steep slopes. Recent floods that have devastated Honduras, China, and Venezuela have been due not so much to storms as to deforestation and poor erosion control. All these countries have the know-how and the political ability to cure the problem; they lack the will.

Conversely, there are many success stories. The air of Los Angeles is better than it used to be. We have more trumpeter swans, whooping cranes, and bald eagles than we did 50 years ago. To set against this, there is a worldwide increase of pollution, erosion, and deforestation. The human death toll from disasters such as Chernobyl, and from more insidious causes such as pesticides (Wright 1990), is substantial.

FEELINGS AND SPIRITS

Environmental experience is important far beyond the realms of conservation and ecology. Marc Berman and coworkers found in several psychological experiments that interaction with nature has "restorative effects on cognitive functioning" (Berman et al. 2008:1207). The low-key but extremely varied stimuli of nature stimulated the mind but did not exhaust or stress the viewers. Interaction with nature both improved performance on cognitive tasks *and* relaxed the subjects just as a good sleep would do. Humans are wired to interact with nature, so these results are unsurprising, though delightful.

We cannot go farther in dealing with the world environmental crisis unless we have a far better understanding of what makes people act. This understanding has to be conformable with the known facts of psychology and with the known cross-cultural data that anthropologists have recorded (including the problems of the desperate and powerless). With this knowledge, we can,

hopefully, make the real dangers more clear, divert attention from more trivial problems, and develop institutions that will actually protect life and livelihood.[5]

Humans are not mere bundles of physical needs. What confronts the nonhuman environment is a world of individuals: total persons, with all their
fears, joys, sorrows, and visions. Resource use is mediated by needs for fuel
and for poetry, for calories and music—the old cry for "bread and roses."
Humans are both economic and spiritual animals. Antonio and Hannah
Damasio have shown that humans cannot separate emotion from cognition
(Damasio 1994). We are creatures of both cognition and emotion, and we
must integrate both in making decisions. Moreover, humans are creatures of
conflict, love, art, anxiety, and all manner of other things.

Social scientists have constructed the most wildly disparate models of
human nature. Some hold the idea that humans are naturally savage killers
and sex maniacs under a thin veneer of civilization. (This view, currently
espoused by certain "Darwinians," goes back to Calvinist religious views of
humans as born sinners; it has no Darwinian validity.) Others maintain the
hopeful dreams of humans as loving, caring, and spontaneously good unless
corrupted by evil society. This view was powerfully stated by the Chinese
philosopher Mencius in the 4th century BC; he explicitly related it to the
environment (Mencius 1970). Some see humans as powerful and tough; others see them as weak and unhappy. Some see them as rational calculators,
dedicated to methodically maximizing their self-interest; others see them as
creatures of culture or of emotion, blown by every passing wind. These various views have their own moral and political implications.

Obviously, a view of humans as evil, violent, and sex-crazed will lead to a
different law code than a view of humans as naturally helping, peaceable, and
caring.

Common experience suggests that people are people: competing or coop-
erating, helping or harming, and often fighting their loved ones instead of
their enemies. People can work terribly hard for the thinnest rewards, and
then be lazy when even a small effort would win much. People (especially—
but not only—teenage people) can be fiercely independent one minute, and
abjectly conformist or dependent the next.

People are social loners, or unsocial socializers. Studies of apes, our nearest relatives, disclose somewhat loose and amorphous social orders. Moreover,
even the social order of modern chimpanzee or gorilla troops may have
evolved somewhat since they branched off from our common ancestral line.
We humans have evolved our own sociability very fast. We are not perfect at
it. All of us are torn, at least sometimes, by the conflicts between maintaining control of our lives and maintaining our social ties. We want to be

independent, yet we want to be socially connected. This inner conflict does not occur in bees or termites, or even in most social mammals. Humans can be "lone wolves," but wolves rarely are.

People do not learn about the world by deductive logic, or even by individual observation. They learn by interacting with actual things (cf. Louv 2005). They interact, think about what they have experienced, apply logical deductions to it, and test the deductions, thus experiencing things anew. They continue the cycle indefinitely. This cycle has produced the thousands of different worldviews that are now being called "traditional ecological knowledge," as well as contemporary science and other less traditional views. As an anthropologist, my area of expertise lies in the traditional knowledge systems—how they form and how they change. I hope that a lifetime spent in dealing with such systems can provide some useful insights into the wider, more complex, and sometimes more troubling systems of thought that are emerging today.

One vexing question about innate human morality concerns what E. O. Wilson called "biophilia" (Wilson 1984). How much do we love nature? Wilson pointed to the universal tendency of children to love animals and flowers, and the general fondness among adults for landscape paintings and gardens. Wilson points out (and see also Orians and Heerwagen 1992) that people everywhere enjoy savannah-like vistas, and feel awe and reverence for mountains, fast-flowing water, and the like.

There is a great deal of evidence for E. O. Wilson's "biophilia" hypothesis (Wilson 1984): People just plain like animals. We have evolved with them—hunting them, using them as companions, and watching them. Thus we remain fascinated with them. Pet-keeping is the most universal and striking form of this. Furry animals are especially popular, but anything will do. Everywhere in the world, people make pets of just about every mammal they can handle, and of the tamer birds and other creatures as well. Iguanas, alligators, pythons, and other unlikely creatures have countless devotees. Conservationists can count on people being fascinated with "charismatic megafauna," but people do not stop there; they are lured by tropical fish and corals, sponges, and even worms. People show a strong and universal preference for mammals and birds. This causes great annoyance to conservationists trying to make a case for insects and microorganisms; however, even a small amount of public education seems to make people care about almost any creature. Bats have been redeemed in recent years. Snakes are instinctively feared by most primates, probably including humans—but they have become popular pets anyway.

People learn to love their usual environments. Anthropologists report that Inuit love the Arctic, Plains Indians loved their bare grasslands, Bedouin love the desert. And all biophilia genes are monumentally underexpressed in the rising generations worldwide. The young of today are city mice (Louv 2005). Hunting, fishing, visits to national parks and forests, and outdoor recreation in general have been declining at a rate of one percent a year since the 1980s, for an overall decline of perhaps 25 percent (Biello 2008). We are losing our concern for nature, and with it our hope of saving the environment. People work to save what they love, not to preserve an uncertain and personally irrelevant "future" (Anderson 1996). By the time pollution and overuse have become obtrusive enough in their own right to force concern, it is too late.

I suspect that Wilson is right, but that biophilia is less strong than a genetically programmed tendency to adapt to *any* surroundings, especially ones that are prestigious or culturally approved. Children still like squirrels, but the prestige of electronic gadgets drives squirrels out of mind. Ironically, children today (my own grandchildren included) often love watching animals on television more than they love being out and about in the real world.

Caring is what matters. Loving nature is critical. Responsibility can only come from actual caring and concern. But it seems that we have the care. Where we fail is in uniting to act on it.

AND SO, TO THIS BOOK'S ARGUMENT: LACK OF SOLIDARITY

The Tragedy of the Commons, made famous by Garrett Hardin (1968; see next chapter), provides a critically important and insightful explanation, but only a partial one.

It serves, however, to alert us to a wider problem. The Tragedy of the Commons happens only when people do not get together to cooperate in working out ways to manage a resource. In all other environmental tragedies, the same is found: The parties most affected fail to join together to act. Under such circumstances, we cannot expect people to act for long-term or wide interests.

I argue at length in this book that the biggest problem facing the environment today is the lack of unity among those who are most immediately dependent on, or concerned with, conservation and sane management. The political powers successfully use racism, ethnic hatreds, economic interests, and other classic devices to carry out the old Roman program of *divide et impera*, "divide and rule." It follows that the environmental agenda advances insofar as it addresses technological and biological issues, political concerns (chiefly the pyramiding of power), and above all the question of solidarity.

Overcoming these problems requires leadership by those who are willing to sacrifice themselves out of a sense of social responsibility. Even in the United

States, environmentalists have been threatened and sometimes murdered. In the Third World, the situation is far worse. Chico Mendez, the rubber tapper who was assassinated in Brazil while trying to save the forests, is only the best-known of thousands of men and women who have been murdered for their environmentalist activities. Thousands more have been intimidated, fired, harassed, or threatened. The "tree-huggers" (Chipko) of India have become a byword for foolishly romantic environmentalism; few Americans seem to realize that the Chipko movement is an incredibly heroic one (Guha 1993). It pits villagers trying to save their forests—which are valuable as well as sacred—against the Indian army and local landlords. Far from wishy-washy romanticism, tree-hugging was a desperate act of self-defense.

Further experiences, as well as perceptive studies like William Ascher's *How Governments Waste Natural Resources* (1999) and Arun Agrawal's *Environmentality* (2005), have shown that complexity is the rule, not the exception, in cases of environmental conflict. In all well-studied cases of this (so far as I know), the problem is not just rational material self-interest. There is always an emotional and political dimension. Usually it involves government. Very often, group hate is the real underlying cause. Often, however, it is simply bureaucratic irresponsibility, power-maintenance (Scott 1998), and inertia. Sometimes, sadly, it is excessive preservationist zeal carried to the point of alienating local people and their global allies.

This wide problem of passion or irresponsibility gets too little attention. Excellent textbooks of environmental anthropology, such as Daniel Bates' *Human Adaptive Strategies* (2005) and Monique Borgerhoff Mulder and Peter Coppolillo's *Conservation* (2005), make a default assumption of individual rationality. As anthropologists, the authors know about culture and about emotion, but these become residual categories (at best) for them. To them, as to almost all environmental economists and many other observers, humans are rational individualists.

Evolutionarily, we know that emotions, moods, reactions, unconscious quasi-instinctive mechanisms, and other back-brain equipment evolved long before rational thought. George E. Marcus (2002) gives a particularly good discussion of this and its political relevance. We know that, as David Hume so well and so memorably put it, "reason is, and ought only to be the slave of the passions, and can never pretend to any other office than to serve and obey them" (Hume 1969 [1739–1740]).

My sense is that most of the landscapes that have been saved in the world have been saved for esthetic and recreational reasons—as national parks and the like. The same was, until recently, true of wild animals; they have been saved for tourism or for sport hunting. We can perhaps do better than saving animals for such uses, but we surely cannot do better than saving them because we want them. Saving them out of fear or compulsion is not a long-term solution. It simply does not work.

Among more managed landscapes, most are saved because they produce multiple benefits, as opposed to one benefit only. Soil and water conservation, mixed farming, sustainable forestry, and coastal wetlands protection provide diverse examples. We save soil because it is basic to all farming. We save forests, when we do, because of the needs of multiple-use management. (This is not true of intensively managed tree farms, but such systems are not true forests and may prove to be unsustainable.)

There is a close link between human rights issues and environmental abuses. These have often been separated, and even seen as opposed—especially when "human rights" are defined to include rights to unchecked use of resources for immediate economic purposes. The rich nations draw shrimp, fish, energy resources, forest resources, and other goods from the poor nations, but leave the poor stuck with the environmental costs. The rich also affect the poor by producing global warming—directly, by burning fossil fuel, or indirectly, by buying products from the poor nations that cause deforestation in the latter. Thus the poor pay for the sins of the rich (Srinivasan et al. 2008). Often, the rich force particular kinds of "economic development" on the poor at gunpoint; extraction of oil and minerals has notoriously been this sort of bargain in the past (Humphreys et al. 2007). For that matter, extraction of slaves in the longer past is still having ecological effects on Africa.

Lack of solidarity thus involves economic injustice, which usually follows from hatred and intolerance. This then becomes *the* problem for modern humanity. It is significant that we treat malaria, tuberculosis, and even depression and schizophrenia, but it never occurs to anyone to treat insane hatred of other groups. We think it is "normal." In the twentieth century, it was a norm, but so was malaria in the nineteenth. Hitler, Pol Pot, and their kin defined that mercifully-vanished century, in which war and genocide killed almost 200 million people (Rummel 1998) and caused a proportionate amount of environmental destruction.

I find terms such as "greed," "modern technology," and "capitalism" to be unhelpful. "Greed" is an outrageous word anyway. If I want a fish, it's "my legitimate subsistence need"; if you want a fish, it's "greed." Most environmental damage is still done in the production of staple foods and fibers, and most of the people who are destroying the environment are desperate small farmers, fishers, and forest users in impoverished countries; do we speak of "greed" in their cases? We should be thinking in terms of *solutions*. When we do, we find that materialist explanations rapidly fail.

The abuse of the environment as part of the abuse of power is no new thing. Long before capitalism, feudal and imperial regimes that got over-addicted to cattle or sheep overgrazed and overcut as badly as anyone since. The Roman Empire and the Chinese state were notorious offenders thousands of years ago, as their own home-grown social critics pointed out (cf. Ponting 1991).

Yet, power is never total. Even the Communist dictatorships of Eastern Europe—the most systematically totalitarian states in history—fell in the end. In fact, they fell rather quickly, as history goes. The united power of the people is vindicated. It could end the rule of giant primary-production corporations and restore reasonable environment use, if people were united.

One cannot simply say that we have to "cut down on consumption," or, conversely, that we have to continue destroying resources "to benefit the poor." Consumption in the sense of resource destruction is not the same thing as consumption in the sense of actual benefits to individuals. One has to concentrate on the actual human-environment interactions involved, if one is to understand the realities of resource use. Cutting down on consumption, for instance, as advocated by many environmentalists, makes sense only to the 10 percent or so of the world's population who can really do that; 75 percent of humanity remains locked in poverty. Even in the United States, 20 percent of the population, and over 30 percent of the children, live below the official poverty line.

Moreover, today, if Americans cut down on, say, consumption of oil, all that accomplishes is making oil cheaper, leading to more use in developing nations. There are economic reasons why one might want that, but it does not help the environment.

Ethical philosophy has tended to speak of "individuals" or of institutions, and thus to collapse into what seems to me a sterile non-dialogue between individualist and "communitarian" writers. This non-meeting of the minds often surfaces as a war of words between libertarians and centralists.

Recognition of the importance of others is the necessary moral grounding for appreciation of diversity—for tolerance, for preserving biodiversity, and for appreciating other people and other cultures. This is basic to solidarity, and second only to it as an immediate need. From monocrop agriculture to ethnic cleansing, the contemporary world has set its face resolutely against diversity of any kind. Standardization, mindless conformity, intolerant hate, and mutual jealousy have combined to produce a worldwide cultural trend hostile to all difference, all deviance, and all variety. This is, at present, the most extreme and direct threat to the environment, as it is to human life and to the human spirit.

The highest environmental and human rights goals cannot be fully achieved, but they are *process goals*: goals so desirable that any progress toward them is worth the effort. As argued by Rene Dubos in *Mirage of Health* (1959), "perfect" health is unachievable in this imperfect world. However, better health is almost always better than worse health. Thus, we have to progress toward perfect health. We will never get there, but every bit of progress in the right direction is pure gain. Similarly, we need all the environmental improvement we can get. Tradeoffs, such as managed productive forests vs. protected old-growth, must be dealt with as such: tradeoffs in an economy of environmental goods.

One way or another, we have to manage to achieve solidarity among environmentalists who want to save old growth, loggers who need the forests but need to manage them sustainably, and everyone else too, since we all need forests as carbon sinks and oxygen generators.

THE REAL CURES

The cures to these problems must fall into comparable broad classes:

New economic and political institutions, using what we know of ecological management.

Moral institutions, new or revived, to cope with the problems of motivation and solidarity. Especially necessary are caring and concern for the environment, and consequent sense of responsibility. This can follow only from solidarity and unity among people—and probably also a sense of solidarity with plants and animals. Traditional communities see their living surroundings as part of their social universe, and generally conserve resources to the extent that they can make such social inclusion meaningful.

Education. One major problem for building solidarity is ignorance. Most people still fail to realize how bad the situation is and how bad it might get. Nor do people know how to respond, when they do realize. A great deal of this ignorance is due to willful misinformation spread by polluting interests, but much is innocent. Many environmentalists get discouraged and lose hope simply because they fail to see the enormous base of technological and institutional wisdom we have. Some environmentalists even seem to have a knee-jerk fear of technology that is as disturbing as an unqualified reliance on it can be.

Ignorance of solutions, especially win-win solutions for problems of this sort, is especially a constraint on action. Many people come to believe that solutions to environmental problems are extremely expensive. This is partly because the farther fringes of the environmental movement have made uncompromising claims, based on esthetics or ideology. Some go so far as to argue that the world economic system, or even all of western civilization, must be remade from the ground up. Indeed, if we wanted to save every flower and tree, such would be the case. Most do not want to save that much. We can save a very great deal at relatively little expense, and that message should reach a wider audience than it currently does. The world currently spends on armaments enough money to put most environments in fine shape.

However, it is also true that there really are no easy answers. Should a resource simply be exhausted, and the money invested elsewhere? Should it be conserved? At what level? How can a system deal with the sacrifices of present-day workers? Should society as a whole shoulder the costs, by compensating workers for giving up some or all of their livelihood? Obviously, no two cases are quite alike, and general principles of equity and precaution may be difficult to apply.

We need a rebirth of horizontal ties, but no longer the smothering ties of the small, isolated communities of decades ago; the need, today, is ties between different people all over the world.

People define themselves, their lives, and their economic strategies with reference to families, communities, and societies. They define themselves, their worlds, and their concerns through such interactions; the interactions, in turn, define the persons. They discuss, restrain each other, complain, and constantly check their ideas and actions against others' opinions. In the real world, all resource management is *negotiated*. It is a matter of practice (Nyerges 1997). It emerges from the interaction of people and environments.[6] It often emerges from conflict, as political ecologists know (Sheridan 1995; Stonich 1993), and the sides are not always as clear as we think. Even the heroic Chipko tree-huggers of India have their ambiguities (Haripriya Rangan 1996 in Peet and Watts 1996).

Martin Buber's theological conceptualization of "I and Thou" made interaction the basis of theology: the worshipper interacting with God. The same is true of our everyday existence, our "being-in-the-world" (as the philosophers say). For Buber, we define ourselves, we exist, through interaction.

This is certainly true of human life. Learning is, by definition, interaction. It involves the organism interacting with its environment. For humans, learning is almost always in a social and cultural context. (*Culture* is herein defined as "a system of learned behavior patterns shared with a wide social group.") We interact with teachers, students, and peers. Learning is thus a personal thing. Buber tells a story of Rabbi Leib studying with the great 18th-century Hasidic teacher Dov Baer. Rabbi Leib said he went there not to hear about the Torah but to see how the Master tied and untied his shoes.

Anyone can lecture on the Torah; a great teacher is one whose smallest act *is* the Torah (Buber 1991:1:107, 169).

As Buber says, we find all our meanings—from the meaning of a word to the meaning of life—through interaction. Watch a young child learning a new word: She sounds it out, uses it whenever possible, and watches how adults react. Slowly she learns, from their reaction, what are the limits of the word. *Flower* does not include leaves or spots on clothing; *dog* cannot be used to refer to cats or rabbits. Our coming to understand the environment is done this way. Our search for the meaning of life, or whatever meanings we can develop in our lives, is done this way. Contrary to certain postmodernist views, neither individuals nor societies "construct" meanings in a vacuum. Nor are meanings "lying around in things"; we have to create them. We create them through interacting with things and with each other.

People used to learn their natural world through growing up in the environment, living in it, drawing their food from it, laboring in it. Today, most of us (even in the Third World—now primarily urban) see natural environments only on television or on rare visits to "the country." We are deprived of a vital learning experience.

Thus, a number of theories of human action have arisen recently that are based on interaction, dialogue (or polylogue), and negotiation. Humans not only talk a lot; they create social usages and institutions through talking together. It took a century of improving communications and deteriorating economics before English workers realized they were a "working class," not just separate groups of Midlands craftsmen, Sussex farm workers, and Tyne-side herring-packers (Thompson 1963). Solidarity, and thus collective action, grows through dialogue.

In other words, practice—actual work, actual play, actual social functioning—involves interacting with other people and with the environment. From this interactive practice, people construct institutions. These institutions, in turn, sum up to what we call "society" and "culture" (Bourdieu 1977, 1990). More: Every individual constructs his or her self from these interactions (Mead 1964). Interactive practice is the reality from which concepts like "Society" are abstracted.

This being so, it follows that interactions are all-important. The philosopher Emmanuel Levinas says they are infinitely important (see below). Our interactions are everything to us. Jurgen Habermas has argued for a politics based on this recognition (see esp. Habermas 1989). I see no reason to cover the same ground in much detail. Suffice it to say that Habermas's ideas of communicative ethics and civil society imply grassroots activism, comanagement, empowerment of actual users, and respect for local traditions that work for whatever task they are intended. Most of this book is taken up with arguing for small-scale, local, often traditional institutions as part of (*not* all of) the solution to managing the environment.

Ultimately, we have to see good resource management as an absolute unquestioned need, like preventing murder and giving emergency medical care.[7] Many more lives are involved.

We also have to love nature, for without love there will never be enough effort.

2

Learning from Others

I began to understand that for the [Native people of western Canada], there is no such place as wilderness. The world is not divided into the natural and the cultural, forever in opposition, wholly different in kind. . . . Nature is *not* there for exploitation or alteration at the whim of humans. . . . The landscape is home.

(Johnson 1997:64)

SOME TRADITIONAL REGULATORS

All traditional societies have managed to live with their resources long enough to become "traditional." To be sure, "tradition" may be very new indeed. Italy has had tomatoes only since about 1700, and Ireland has relied on potatoes as a vital crop since about 1800. But, still, that is long enough for these New World crops to have been worked into resource-management plans.

In general, what we can learn from the others is, first, how to live with nature instead of defying it; second, how to be efficient and economical rather than maximizing throughput; and, third, how to maintain solidarity.

Several people have recently argued that learning from other cultures is "colonialist expropriation." This is correct if the knowledge is copyrighted, or if it is being expropriated for direct profit. However, if we are simply sharing widely known truths for universal benefit, we are doing what all people have done throughout history: learning basic life skills from each other. No culture is pure or isolated; none is self-sufficient in knowledge; none has gone even a few years without learning from its neighbors. In a world where almost everyone eats hamburgers, has a cell phone, and watches rock music on TV, there is little hope for cultural purity. More to the point, attempting cultural purity today would be suicidal. We have to learn from others, because we no longer have the time and resources to experiment and learn by trial and error. All societies have been working on these issues for millennia; probably every society on earth has some knowledge useful to all. If we do not share this knowledge now, we will suffer proportionally.

A huge amount of literature deals with these matters and lessons. Several journals are devoted entirely or in part to this topic. They range from the highly scientific *Ecology and Society* (online) to the spiritual and humanistic *Journal for the Study of Religion, Nature and Culture.* Many scholars have devoted their lives to relevant questions; for example, Fikret Berkes (e.g., 1999), Darrell Posey (1999, 2004), Gary Nabhan (1987, 1998), and Nancy Turner (Turner 2005; Deur and Turner 2005). I have detailed some of the lessons in a number of publications (Anderson 1996, 2005; Anderson and Medina Tzuc 2005; etc.).

Chinese protection of the environment was largely rooted in the cosmological belief system known as *feng-shui* ("wind and water"; see Anderson 1996). This system, basically scientific but greatly amplified by religious and mystical beliefs, was designed to maximize the good influences in the environment and minimize the bad ones. Among the more practical teachings were injunctions to situate one's house on the leeside of a hill, among trees, near permanent water, and above the good cultivated land. Also practical was the belief that a house should face south; this means sun will enter the door in China's rather harsh winters. Practical also, in their time, were the instructions to have a crooked path to one's door (evil influences and bandits prefer to travel in straight lines), to have two door panels, to avoid living in the room opening directly on the front door, and to live near a taller pagoda or church that can beam good influences downward to one's home (and focus a community for mutual protection).

The Chinese carefully protected large feng-shui groves next to villages and temples, for religious reasons but also for use; the groves were managed to provide ready supplies of firewood and building material. The Chinese groves were in immediate proximity to major settlements, not in remote places.

The whole was based on a belief that the breath (qi) of the cosmos flows through all things, including mountains and rocks. Properly tapped, it provides wealth and fortune. Improperly managed, it allows evil chances to come. Good spirits, such as dragons and tree spirits, are also part of the scenery, and their help should be sought. From the outsider's point of view, these are symbolic spirits, emblems of the very real benefits of hills and groves (Anderson 1996).

The Chinese were aware that the environment, broadly defined, needs some kind of long-term defensive managing. The benefits of established groves and forests are very well recognized. So are the benefits of clean, fresh water. The dangers of overhunting and overfishing are well laid out in the foundational Confucian texts of the fourth and fifth centuries BC, and, until recently, some of these teachings were known to every Chinese schoolchild. Everyone realized that damage to the environment, or at least to these elements of it, damaged everyone's interests (Anderson 2001; Anderson and Raphals 2007).

Unfortunately, moral codes are not always effective in this imperfect world, and China's dreadful history of deforestation, water mismanagement, and overuse of plant and animal resources reminds us that the best intentions often do no more than pave the path to a well-known place. China, "land of famine" (Mallory 1926), has all too often faced situations that forced people to do anything they could to survive in the short term. Nor have the rich always seen fit to act morally, even when they did have the option. The Chinese did what they could, although they were far from being able to maintain their environments in ideal shape (Anderson 1988, 1996; Marks 1996).

However, things could have been a very great deal worse, and I am certain they would have been worse without feng-shui and the rest of the system. The Communist government has been carrying out a huge experiment over the past 50 years. They moved quickly to eliminate "feudal superstitions" such as feng-shui, and to "develop" China's resources as fast as possible. The result has been devastating deforestation, overfishing, wholesale extinction of wild animals, soil erosion, and more (Economy 2005; Smil 1984). This is not the automatic result of modern technology; actually, modern technology, applied according to the old moral views *or* according to reasonable market principles, would have saved China. Proposals have been made throughout this century to use modern technology to reforest, to regulate water, to control soil erosion, to conserve game, and to otherwise act according to traditional Chinese values.

Veronica Strang, in her book *Uncommon Ground* (1997), describes the different attitudes of Australian Aboriginals and Anglo-Australian ranchers in northeast Australia. Both groups are devoted to the land. The Aboriginals, however, use every product of it, and are intensely bound to it by long association. The stockmen are concerned only with cattle, a nonnative species. They thus tend to treat the land as something to be converted to an ideal cattle home, and they have a much less deep interest and involvement with its total resource base.

Kat Anderson (Blackburn and Anderson 1993; K. Anderson 2005) finds the same thing in California. Her work with the Native Californian peoples has shown that they draw on the foothill and mountain landscapes for hundreds of different plant and animal resources. Most of the areas in question now produce only a few cows, or small amounts of timber.

In my research in Quintana Roo, I have found the same pattern. The indigenous Maya people use thousands of plant species and hundreds of animals. They know exactly how to get the fullest possible value from every rock, every soil pocket, every bush and tree. They appreciate not only the medicinal and food uses of plants, but the value of plants for wildlife, firewood, soil restoration, string, rat poison, and dozens of other uses. Mexican developers, by contrast, usually see the forests and brushlands as something to be cleared for cattle or rice or some other alien monocrop. All schemes to convert the Yucatan bush to monocrop agriculture or grazing have failed dismally. (Cattle, however, flourish on natural grasslands in the area. Cattle belong on grass. It's conversion that does the damage.)

BROADER THEMES

Most cultural resource management systems are rather stable over time. The Mediterranean–Near Eastern system, with its heavy grazing by sheep and goats, devastated the landscape as early as 5000 BC, but it still goes on. Such grazing can devastate landscapes under some conditions, as it did when it was exported to the New World. The North European system, based on mixed farming and maintenance of forest for timber (and eventually for parks), has persisted for thousands of years and has been transported to North America. China's intensive wheat and rice agriculture, backed up by land management concepts, lasted for thousands of years, until the Communist government abolished it. India and native parts of the Americas have their own stable systems.

Even purely moral principles, without institutional rules of any kind, can be effective in traditional societies. An example of moral conservation is provided by "culturally modified trees" (CMTs to the anthropologist). All over the American Northwest and Inner West, one finds living trees that have had bark or wood harvested from them. In the Northwest, Native peoples still remove bark from many trees. This is done for food (inner bark of pines and some other trees is edible) or for fiber. Especially important for ceremonial purposes as well as for ordinary fiber is the bark of the Western red cedar (*Thuja plicata*). Native people remove only one narrow strip per tree, so that the tree recovers and can be cropped again and again. (On these matters, see, e.g., Turner 2005.) In the case of a rare tree with very valuable bark—as the Western red cedar is in much of the Northwest—this meant self-denial of an almost incredible order. One might have to walk for miles to find several good bark trees. It is worth noting that all the Northwest peoples said a short but deeply felt prayer to the tree before taking the bark, and thanked the tree for providing it (Kirk 1986).

In the Great Basin, there are countless junipers (*Juniperus osteosperma* or *J. occidentalis*), from which a long, straight strip of wood has been removed. This strip was used to make a bow, juniper wood being the best bow wood in that area. Junipers with straight, tough wood are rare; usually the trees grow crooked and knotty. These bow trees were studied by Philip Wilke (1988). He found that a good straight juniper was a resource known over hundreds of square miles. Yet, no more staves were taken than the tree could support—about one per branch per 20 years. Countless individuals, hunting alone in the bush, would pass by the tree in that time, and would be quite aware of its qualities; yet they forbore to harvest it. This level of moral restraint certainly disproves a great deal of the recent literature in resource economics!

I have observed the same thing in the Northwest, where yew trees provided bow staves. Old yews have a few long, straight cuts in them, where staves were removed. Yet the trees were not overharvested. I suspect the same was true in old Europe. Many ancient yew trees there date back to the days of archery. They must have been cropped for bow staves like the ones in the Northwest. At Blarney Castle (of Blarney Stone fame), for instance, I saw dozens of yews hundreds of years old. I suspect they were planted as a bow stave supply source.

Many Native people who passed up a good bow staff would have had no serious hope of ever getting a bow staff off that tree. At best, they would have to wait for years. But they passed up the tree anyway—not because of their long-term self-interest, but because of internalized social rules.

A similar case of amazing self-enforced plant management comes from Peru. Above the Colca Valley, I observed a tundra bush called *yareta* growing at 13,000–14,000 feet above sea level. It is the preferred local firewood source. It grows about half an inch a year (my measurement) and is not abundant. Yet it is cropped sustainably. Bushes that must be hundreds of years old, given their size, are still producing firewood. A stick is taken, and the bush is left to recover. This has to depend entirely on self-policing; no one could enforce it on others, and no one tries.

Sacred forests are inviolate in many parts of the world (Leslie Sponsel, pers. comm, 2007). A vast area of religiously-constructed traditional resource management extends through Siberia, Japan, Mongolia, and Tibet. Mongols, Tuvans, and their neighbors have a powerful conservation ethic—again, largely wise-use, but with true preservation of many sacred sites (see especially the amazing collection of Tuvan material by Kenin-Lopsan, 1997; for the Mongols, Humphrey and Onon 1996; Metzo 2005; Roux 1966). Tibetans preserve whole mountains (T. Huber 1999) as well as many smaller sites. They carefully avoided overhunting game, although they depended on it; when the Chinese finally occupied Tibet en masse after 1959, a game holocaust took place. The previously abundant and easily shot game was exterminated anywhere the Chinese went.

Tibeto-Burman peoples of west China are equally good managers. Core areas were safeguarded by the Akha and other Tibeto-Burman groups of Yunnan, for instance (Goullart 1955; Ayoe Wang, ongoing research, reported to me in working papers, 2005–2007). Usually, sacred groves are managed for use; their trees can be cut for major public projects, their shade is valued, their fruit and fibers are extracted. They are not dead space but woodlots protected by sanctity. I have observed this repeatedly with feng-shui groves in China and sacred groves from Madagascar to South America. Sacred groves elsewhere have been managed similarly.

Traditional tree cropping in tropical areas preserves many, or even most, of the benefits of natural forests. Areca nut (a palm nut with a stimulant alkaloid, used for chewing for a caffeine-like lift) is cultivated very widely in south and southeast Asia. The small palm grows in the shade, and thus a somewhat diverse and multistory forest or garden is needed. Areca cultivation in south India has gone on for 2000 years or so, with 90 percent of local forest birds happily inhabiting the areca groves (Ranganathan et al. 2008). I have seen similar systems all over south and southeast Asia, with areca as one component of mixed tree-crop systems.

The conservation of fish is another case in point. On the Northwest coast, traditional fishing peoples actively conserve fish and even stock streams (the Nuu-Chah-Nulth of Vancouver Island did, for example). This is not preservation for preservation's sake, but careful managing. The stocks are invariably fished hard, but usually not hard enough to deplete them. Native societies in this area conserved nature for religious reasons (Anderson 1996; Deur and Turner 2005; Turner 2005). These nations believed—and many people still believe—that every tree, mountain, and animal had its indwelling and powerful spirit. These spirits were real persons, who could communicate with humans and were part of a wider society in which humans were only one class of participants. Thus, humans had to make humble requests to trees for their bark, to berry bushes for their fruit, and to animals for their lives. Such requests were taken very seriously indeed, as I know from personal observation. Taking too many animals, or taking them without the proper rituals, alienated the animal spirits, and guaranteed failure in the next hunt. Obviously, this belief stems from repeated observation of the effects of overhunting—but, where the modern biologist speaks of simple overkill, the indigenous explanation is that the spirits are angered by disrespect, and take their leave.

On the Northwest coast, as elsewhere, people are not always perfectly moral. I heard of cases in which individuals "robbed creeks" (took all the salmon, permanently wiping out the run, leaving none for others) or otherwise blatantly violated the moral code. Such persons were shunned and criticized, and could wind up in real trouble. In traditional times, with traditional technology, that threat was probably quite adequate to protect the resource; at any rate, even the tiniest streams had salmon in them when the whites came. With today's technology and market system, and with the wider society crowding in, the moral code needs some shoring up.

The Maya of Yucatan have a rather similar, and distantly related, belief system (Anderson 1996, 2005; Anderson and Medina Tzuc 2005). Less spiritual (at least today), it merely holds that a properly socialized person will not take too much. I have seen Maya, including my coauthor Felix Medina Tzuc, carefully move bugs and ants out of the way of foot traffic. However, older persons still believe in the Lords of the Forest, the gods who punish overexploiters. Every aspect of cultivation and hunting is religiously—and often ritually—represented. This system allowed the Maya to live and flourish in a difficult environment, and build one of the world's great original civilizations there. (Its much-vaunted "collapse" affected only the central lowlands, not the much more ecologically difficult north. It was due largely to drought and warfare, not—contra Diamond 2005—to simple ecological overshoot; see Demarest 2004, Gill 2000, Webster 2002.)

The Maya used to hold a ceremony after taking 13 deer; the 13th deer had to be shared out to the community at a large feast with a religious leader officiating. The religious leader would offer the appropriate thanks, and prayers for more luck. Maya hunters explained to me that the expense and difficulty of holding this ceremony forced successful hunters to stop hunting for long periods of time, taking pressure off the animal populations (Anderson and Medina Tzuc 2005). Since the ceremony is a dead letter, we shall never know how well it worked. There were similar ceremonies for renewing the "luck" of a rifle or shotgun, and for other hunting activities.

Not all traditional groups (peasant villages, tribal societies, long-established fishing communities) have conservationist values. Many do not have the values at all (Hames 2007). This is reported especially from South America, where peoples of the tropical rainforests and dryforests seem lacking in ideas of sustainable management (Alvard 1995; Hames 2007). In some other regions of the world, people have good ideas, but weakly held or weakly enforced (Anderson 1996; Kay and Simmons 2002; Kottak and Costa 1993—the last of these reporting from Madagascar, and I can confirm this from personal research). Obviously, no completely irresponsible way of life could last long enough to become "traditional," so all traditional communities must have some sort of management rules, but the rules can be hopelessly inadequate in the modern world (Kottak and Costa 1993).

Contrary to the myth of the "ecologically noble savage," traditional cultures are usually less than perfect in their resource management. In fact, some do worse than we do, if the measure of success is maintaining a population at some level of adequacy without trashing the environment in the process.

Some (notably Hames 2007; Smith and Wishnie 2000) have argued that traditional management is not "conservation" in the modern sense, because traditional management is for short-term utilitarian reasons. This objection does violence to normal understandings of the word "conservation" in English. We speak of soil conservation, water conservation, range conservation, and biodiversity conservation. None of this involves lockdown protection. It usually involves active, ongoing management. Gifford Pinchot is more broadly ancestral to modern conservation than John Muir. Alternatively, these and others maintain that traditional management is for the present only, while conservation has to consider the future. This is usually not the case. Burning, pruning, building irrigation works, saving and planting seeds, preserving sacred groves, and other practices are clearly future-oriented and involve foregoing resources now.

Traditional peoples are similar. The Native American ecologists Raymond Pierotti and Daniel Wildcat (1999) point out that Native Americans did not "preserve" the land—they managed it, often quite intensively; they kept it healthy and productive, according to their needs, rather than locking it up. Their actions created a great deal of the prairie, forest, and meadow landscapes that European settlers mistakenly considered "natural" and "wilderness" (see K. Anderson 2005 as well as sources mentioned earlier).

Levels of indigenous management, however, vary enormously among traditional and local communities. At one extreme are groups like the Piro and Matsiguenga of Peru and the Hadza of Tanzania, who practice minimal management and little or no conservation; they simply do not think about sustainable use (Alvard 1995; Hames 2007; Terborgh 1999; it should be noted that even the Piro and Matsiguenga farm, and thus conserve seed stocks at least). This is because they have not had to; they live at very low population densities in a very rich environment. Similarly, pioneers in newly colonized environments often exploit the new and unfamiliar land shamelessly and mercilessly, leaving devastation in their wake. This is all too well-known in American history, but it is by no means confined to Anglo-American settlers, and is also attested in Polynesia (see below) and many other places. (This would include Native North America, if it is true that the earliest migrants exterminated the megafauna. This is widely alleged, but there is no real evidence; see, e.g., Krech 1999, a work that most certainly does not idealize Native American resource management!) There is normally a learning curve, as people figure out how to use the environment more sustainably. This may take a year or a century, but it generally happens, except in situations where massive erosion or urbanization renders "return to paradise" impossible.

However, these are extreme cases, and so are the ones described by Jared Diamond in *Collapse* (2005). There is, unfortunately, a large segment of the human-ecology community that maintains that all humans, at all times and places, are as bad as the worst developers and agribusiness landlords of today. If this were true, the human race would have become extinct thousands of years ago. Diamond holds forth some faint hopes for the future in *Collapse,* but if his highly colored accounts of the past were true, there would be no hope; humans would simply be unable to get together. These authors follow Thomas Hobbes, who famously argued that humans were in a state of constant "warre" until they got together to put themselves under command of a king. As everyone since Hobbes has pointed out, if people were really Hobbesian they could not make a social contract, no matter what they wanted. People have to have been social from the beginning. The situation is the same with environmental management.

Fortunately, the Piro and Matsiguenga are extreme. At the other extreme are the groups I have studied: the south Chinese and especially the Tibeto-Burman groups of west China; the Maya of Quintana Roo; and the Native peoples of the Northwest Coast. Of these, the Maya have the most explicit ideology of sustainable use (Anderson 2005; Anderson and Tzuc 2005) and, not surprisingly, the most comprehensive set of sustainable practices. And the classic Maya did not "collapse" in a Diamondian sense. Diamond ignores the facts that only the central Maya areas collapsed, while the other Maya lands did well; that warfare was known to be a part of the story (Demarest 2004); and that drought, not ecological overshoot, was the major ecological stressor (Gill 2000).

What is shared by these peoples is a sense of using everything efficiently—with minimal waste. The systems' throughput is what is necessary for human life and system continuance; there is not the enormous pileup of pollution and waste—both of which are inefficient use of valuable resources—that characterizes the modern world. People in traditional societies use hundreds of species, rather than only a few. The staple food is featured, but hundreds of other species are used. This provides an incentive to keep the whole ecosystem healthy, since so much of it is useful. Thus even quite minor species are valued (and often thought to be religiously important). The contrast with the contemporary world, where only ten crops dominate world trade and all the songbirds are dying out because almost no one cares, is striking. Many other traditional societies are less careful, but at least none is dedicated to maximizing throughput. All at least recognize some economic efficiency, and most plan for a sustainable future. This involves ideology and cosmology as well as economic thinking.

What matters are the many cases where traditional societies manage resources for intensive use and yet still do not destroy them. Some societies have intensively fished vulnerable stocks for thousands of years. Others have farmed tropical rainforests for equally long times, without destroying the forests. Others have grown row crops for millennia without losing all their soil in the process. Even the much-maligned Africans often do very well, except where slaving and colonialism have ruined their management systems (Beinart and McGregor 2003; Zimmerer and Basset 2003). We need to understand this (Snodgrass and Tiedje 2008).

Large game animals are often exterminated; they can easily be hunted out, even by low-density human populations. Fragile island species, too, are very vulnerable, and disappear when people and their commensals (rats, cats, pigs) take over. Normally, however, humans live at higher densities in less vulnerable landscapes, and forage for smaller-return, more easily manageable resources, such as small game, fruit, and seeds. Under such cases, it pays to think of one's society-mates and of the future. Management of such resources is virtually universal.

I grew up in rural Nebraska more than 50 years ago. In the 1920s and 1930s Nebraska filled up and hit an environmental brick wall. Overuse of land and resources led to massive soil erosion. Streams dried up. Game was gone. Thousands of farms failed, and tens of thousands of people were homeless and impoverished.

Conservation had always been known to a few people, but from the 1930s dust bowl, it took off as a concept. The Second World War intervened, bringing Americans together and creating a sudden, desperate need for food and fiber production. The United States government under Franklin D. Roosevelt was acutely conscious of the need. Soil, water, and wildlife conservation was propagated in every school, farm association, and government agency. Agricultural handbooks appeared everywhere.

All this entered local rhetoric. I was taught conservation by everyone from schoolteachers to hunters and fishermen. All but the most intransigent people were conservationists. The "latest scientific findings" and the "advice of leading experts" was the language used. Behavior did not meet ideals, but the land did get saved. Moreover, the people who poached, or who polluted the water, knew—as they had not known before—that they were being bad. They may have been defiant, but they lost self-respect.

For a couple of decades, tree planting, soil erosion control, water resource management, and wildlife increase were dramatically visible. Then government policy turned around. Under Nixon and Reagan, the policy shifted to one highly unfriendly to conservation and to such "minor" benefits as wood-lots and game on the farms. It has become steadily more anti-environment since. Giant agribusiness and chemical corporations have taken over both the farmland and the relevant government branches. As in the 1920s and 1930s, a huge wave of farm failures followed, especially in the mid-1980s. The environment began another downward slide. On a recent visit to Nebraska, I found the biggest change to be the loss of trees. All the old shelterbelts, woodlots, and orchards were gone. The soil was blowing away. The original prairie had been treeless, but the ground had been protected by dense sod. Now, bare fields are the rule. Another dust bowl is developing.

More damaging has been the effect on people. All of a sudden, being an irresponsible, wasteful, game-hogging farmer was a *good* thing. The vast majority of farmers had sold out and gone to the city, there to find a poor living, in all too many cases. The few who remained often acted outrageously. The outrageous ones were admired and became local leaders. Responsibility seemed to have gone by the board.

I was struck by how very much more effective the environmental education of my childhood was than that which we offer now. Of course, education in general has declined. Environmental education, in particular, is incompatible with a system based on cramming for mindless standardized tests. But there are more specific causes. The campaigns of the 1940s and early 1950s appealed to people's immediate, practical interest—to their concern for their very livelihood. They also appealed to common experience; everyone knew the results of overhunting, and after a few years of posters, everyone knew about soil erosion. The campaigns were quite intensive; they saturated the available media. The campaigns were phrased in direct, clear, easily understood language, and they reached even the least-educated farmers. Finally, the campaigns had a broad base of support and a broad base of agreement.

By comparison, the contemporary environmentalist scene seems disturbingly fragmented and narrowly based. The conservation organizations too often see rural people as The Enemy. Environmentalists have developed a shoddy reputation as too urban, too elitist, and paying too little attention to farmers and other rural workers (White 1996). Many urban environmentalists seem more concerned with NIMBY (not in my backyard) issues or with recreation than with saving the human life-support system.

The small farms of the Midwest, idealized by Aldo Leopold (1949) among others, may not have been wilderness, but they were infinitely preferable ecologically to the vast pesticide-saturated monocrop expanses of today. The latter are obviously unsustainable and are a disaster even in the short run, and everyone knows it. We could easily go back to far better systems. With modern advances in crop varieties, orchard culture, multicropping, polyculture, organic farming, integrated pest control, and much more, we would actually increase production, not decrease it.

If we could do it in the 1930s and 1940s, we certainly can do it now. We lack only the will—the desire and the solidarity.

HOW THEY DO IT: INSTITUTIONS

In short, far from being innate in the human animal, *the current destruction of the environment is an extremely new thing.* Even the enormous transformation of the environment that came with farming did not really trash the world on a huge scale until the 19th and especially the 20th centuries. Farmers found ways to accommodate. This is true even in the Mediterranean, where both intensive farming and a particularly negative attitude toward nature and the wild go back for thousands of years. A. Grove and Oliver Rackham (2001) devoted a large and beautiful book to environmentally sound and sustainable farming practices in that region.

Something has really gone wrong in the last two centuries, and the rest of this book will elaborate on that.

Whether a common property management group is old or new, it must have certain characteristics. Logically, it must have enough cohesion to set rules and enforce them, and enough sophistication to monitor levels of resource use well enough to identify overuse. Elinor Ostrom and her associates have carried out research on a large number of management groups, both traditional and recent, and they identify certain design features as the minimal shared features (Ostrom 1990; McCay and Jentoft 1996). These include clear boundaries, reasonable rules, collective-choice arrangements, monitoring, sanctions that accord with the seriousness of the offence, conflict resolution, reasonable top/bottom relationships, and organization.

This has allowed conservation and management of everything from Maya forests to lobster stocks in modern Maine (Acheson 1998, 2006). When traditional local management regimes fail, the cause is often outside interference rather than local incompetence (Agrawal 2005; Anderson 1987; McCay 1997).

Without a preexisting order, no such system could arise. Even with a preexisting order, such a system would be exceedingly hard to create. Particularly difficult would be cases in which powerful individuals had a vested interest at stake. The failure of grazing reform in the United States, as well as Ostrom's case of the failure of water regulation in the Mojave River basin (Ostrom 1990), demonstrate this point. Indeed, we would expect management only when immediate pressure threatens the very existence of a more or less homogeneous community with good knowledge of its resource base. Yet Ostrom and her co-workers have demonstrated that successful management groups arise exceedingly commonly, in both traditional and modern settings, even in some situations that would appear quite uncongenial to them. In my personal experience, indeed, only quite powerful political forces can stop the formation of such groups or institutions (Anderson 1996).

Some success stories of traditional management are worthy of particular note. Forests and groves have been saved in diverse parts of the world. Particularly interesting is Conrad Totman's book *The Green Archipelago* (1989; see also Totman 1995), which chronicles the reforestation of Japan under the Tokugawa shoguns. They took a no-nonsense approach: Villages and estates had to reforest, or else lose their lands and a few lives into the bargain. The methods were Draconian, but the success was astonishing; a mostly deforested archipelago was 90 percent covered with some sort of forest by 1868, when the Tokugawas fell.

Whole systems, sustainable or nearly so over long periods, characterize the agriculture of south China, old Polynesia, and much of the Native New World. Polynesia is a particularly revealing case. Archaeology shows that early settlement was carried out with typical pioneering thoughtlessness. Sometimes this permanently ruined whole islands, as in the case of Easter Island. Often, resources (especially flightless edible birds) were depleted, and the human population crashed. However, on larger islands, people learned to manage the land well (Kirch 1994, 1997, 2007). Extremely fine-tuned systems arose, maintained by fertility control and by strict tabooing of overuse. In fact, the word "taboo" is Polynesian, and referred originally to religiously sanctioned restrictions against taking unripe fruit, scarce timber, and other resources that needed to be conserved. Raymond Firth, in his classic studies of Tikopia, described the system as it worked there. He saw it collapse as missionaries converted the people away from the old religion. Famine and disaster resulted (Firth 1959).

HOW THEY DO IT: ETHICS

The commons management literature has only begun to probe deeper issues of ethics and emotion. Many, though far from all, traditional peoples have strong environmental ethics (Bierhorst 1994; Callicott 1994). These ethics serve as starting points for us, but they are too local—too closely identified with particular local groups—for the global mess we are in. We have nuclear power plants, global biodiversity loss, worldwide dispersal of mercury, and other worries mercifully unknown to the traditional societies.

Thus, *we should be fully documenting local ecological knowledge, and, above all, management systems.* This sort of ethnographic documentation is, after all, what anthropologists do best. It would provide knowledge of how much the local people are managing their area, and how they could potentially manage it if full advantage was taken of their more sustainable or conservationist strategies.

Successful strategies should obviously be encouraged. This would solve many local problems. It would, among other things, get local communities and environmentalists on the same page whenever possible, instead of setting them against each other. It would also fit with the imperative moral need to allow self-determination to local communities. We cannot go back to the days when every community on earth was a totally independent little world, usually at war with any neighbors, but we have certainly gone too far in the other direction. The dominance of giant corporations and their captive states is not a happy outcome for anyone, least of all for local communities dependent on their relations with their landscapes.

This begins to look like the start of a wider world morality. Traditional ideas of world as garden, world as sacred space, and world as community of beings need to be dragged from the realm of mystification.

We have to take traditional indigenous morals and social theories seriously. This may involve not only taking their social knowledge into account, but also a "willing suspension of disbelief" (as Coleridge said of reading novels) about their religious beliefs (Nadasdy 2007). So far, to my knowledge, there is only one philosophically adequate account of a traditional morality system, and it is more than 50 years old: *Hopi Ethics,* written by the utilitarian philosopher Richard Brandt (1954). Many partial accounts of indigenous ethics, especially land ethics, have appeared since, but Brandt's is still the only full-length study by a philosopher who actually did participant observation. It disclosed exactly what we need: a sophisticated ethical system acutely conscious of society, emotion, and environment, including what we may call the personhood of all beings. Many excellent books by anthropologists and ethicists have covered some of the same ground since (e.g., Berkes 1999; Milton 2002; Turner 2005). We need more details. Above all, we need these books and Brandt's and many more, to save us from the very partial and incomplete accounts of local knowledge and ethics supplied by the political ecology literature.

Political ecologists have been acutely aware of the plight of local communities, but have generally been slow to use local ethical and moral philosophy to address the issue (one shining exception is Agrawal 2005). More usually, they ground themselves in a folk-Kantian ethic like that described above, or in a Foucaultian skepticism that rejects or relativizes all moral discussion.

The disadvantages of a traditional community and its institutions are that change is difficult to effect, and the institutions are suited to traditional conditions that are subject to rapid change in the modern world. Conservation rules adequate to a population without guns or gasoline are inadequate once those commodities become common. Population growth makes traditional land and water allocation methods inadequate. Farming methods must intensify to feed more people and produce for markets.

Traditional communities are far more receptive to change than popular opinion holds. This is good in that they can often bend their institutional framework to deal with new problems, but bad insofar as new methods of using the environment cause a rapid escalation in those same problems. (If the above obvious points need citation, I can certainly attest them from my personal experience in the field.)

The morality of a perfect world would be a basic one of concerned citizenship: Care, care for, care about. It would be an ethic that could be accommodated easily in any religion or political ideology, without interfering too much with the process of living.

HOW THEY DO IT: EMOTIONS AND SOLIDARITY

Religion is very often the way of maintaining such systems (Anderson 1996; Berkes 1999; Snodgrass and Tiedje 2008; Sponsel 2001a, 2001b). The word "religion," when applied to traditional societies, usually covers more than it does when applied to modern industrial ones. Anthropologists and other scholars of religion tend to use the term to cover all beliefs about the cosmos, including a good deal of what they call "science" or "cultural belief" when they talk about modern worlds. "Religion," in this broad sense, thus covers most environmental regulation. The reason for such broad usage is that environmental management is almost always regarded as a divine charge, enforced by supernatural beings. Most cultures, traditional or other, see the universe as created by divine beings, who want it to be maintained and managed for general welfare. The "stewardship" charge in Chapter 2 of Genesis in the Bible has analogues in almost every culture. Morality is supernaturally given and supernaturally enforced; proper behavior toward the environment is part of general morality.

The view that religion applies only to worship activities (typically on one morning a week, at most), and that other aspects of life are "secular" and controlled by material self-interest, has a strange and convoluted history. Suffice it to say that this view is not found even in most contemporary societies, let alone traditional ones. Religion as carrier of general morality, including—or even *especially*—morality toward the nonhuman world, is the norm everywhere. The advantage here is that supernatural sanctions are available to enforce ecological common sense. Taking too many animals or cutting too many trees brings down the wrath of the spirits. This is not always adequate (I have seen it fail in Mexico), but it is much better than nothing, and it seems to have been instrumental in preserving such wildlife and forest as remains in much of the world.

Religion maintains itself not only through "scaring hell out of people" but also through rewarding them with festivals and good times (Durkheim 1995). For social entertainment, nothing beats a good festival, be it Christmas, Passover, the Niman Kachina, the end of the Fast of Ramadan, or the Great Yam Feast. Every functioning society has at least one major festival that brings everyone together, provides lots of fun, and teaches basic values. Every functioning society has at least one annual festival that involves children to a very important degree and that teaches as it entertains. In the United States, Christmas serves this function. No longer a strictly Christian festival, it has become America's way of modeling generosity and family solidarity for the young.

I suspect that the more successful the society, the more and bigger its festivals, although no one seems to have done the statistics on this. Certainly my experiences in China support the idea. Chinese festivals are frequent, noisy, and delightful. China is famous for its family and community solidarity and mutual aid.

Even without festivals, a traditional group can draw on a long-accumulating reservoir of social solidarity and good feelings. Also, the people in a small traditional group are bound together by many ties. They have grown up together, worked together, played together, and often fought together to resist enemies. They have had to deal with common problems. They have intermarried, raised children, and gotten together to teach those children. An American living in a Chinese village is soon struck by the fact that any older person can discipline any child. Values are shared, children are part of the whole community, and all responsible persons are expected to help in teaching the children proper behavior.

A number of recent reviews of institutions that preserve nature emphasize the value of religious prescriptions and proscriptions, as well as of religious ethics in general (Berkes 1999; Berkes et al. 2000; Colding and Folke 2001). Religion not only teaches care for the environment; it successfully motivates people to follow the rules, even when no game warden is near.

Religion, in most traditional societies, is more encompassing than it is for many modern people. (In fact, traditional societies rarely use words equivalent to "religion," because spirituality informs all things. I use the word for convenience here.)

All of life, and especially all interaction with natural environments, is powerfully informed by concern for spiritual matters and for what we would call "supernaturals" (but which local people consider perfectly real parts of their environments). There is considerable range in this—it is not true that all Native Americans are "spiritual people"—but the generalization holds. Religion is much more involved in immanent decisions about planting, cultivating, hunting, and gathering, even among my Maya friends in Yucatan, who are among the most hard-headed, pragmatic, and secular people I know; they may never darken a church door, but they still pay great respect to the powers of the fields and forests. Many of my Chinese and Northwest coast friends were much more deeply involved in concerns of the powers of the natural world.

First and foremost, religion creates, maintains, invigorates, and empowers solidarity (as definitively argued by Emile Durkheim, 1995). This is what environmentalists and concerned citizens most lack and most need today.

Second, religion constructs and maintains a social universe wider than one's immediate circle. For almost all non-Western cultures and more than a few Western ones, it maintains a social universe wider than humanity. St. Francis of Assisi stands out within Christianity for arguing personhood for the sun, for animals, and for plants; in most societies no one would ever think otherwise. Most societies worldwide at least regard animals as "other-than-human persons" and treat them accordingly—not as well as one treats one's family, but not with the callous indifference seen in modern industrial farming systems.

Third, religion not only supports and maintains ethical systems, but it enforces them by creating a conscience in the believer. At its least subtle, religion can "scare hell" out of the believer, but usually it operates by higher motivation. People really do want to do right. Also, religion motivates enforcement. Those who do not do right are shunned, and ostracism is fatal in many subsistence-level communities. It is notoriously hard to enforce rules in America now, because we are usually "too polite"—read: not courageous enough in our convictions—to call out a litterer or vandal or poacher. Religion provides both motivation and ready excuse to whistle-blowers.

Fourth, religion gives people hope and strength. This is a cliché, but only because it is so obviously true. The point here is that people without hope or strength simply do not care for the environment, or indeed take much concern for long-term, wide-flung goals of any kind (Bandura 1982).

Fifth, religion often teaches and creates real love. Religions vary greatly in this—Christianity's uncompromising message of love is unusual, if not unique—but love and care, including love and care for nature, is generally at least implied. Much more common, and highly interesting, is the wider emotional engagement that tends to get rather weakly translated as "spiritual power" or "holy power" or some such term. Northwest coast Native peoples express this in their art—the amazing transformed animal figures on their totem poles and housefronts.

Sixth, religion often motivates learning more about the environment. Again, engagement with other-than-human persons implies a far richer and more interested interaction than what we typically find even in rural settings in the (post)modern world.

Seventh, besides the festivals, religion usually uses arts—from painting to dance, from music to landscape modification—to communicate its emotional messages and get everyone involved (as Durkheim 1995 pointed out for ceremonies).

One could go on, but the package is clear by now: Religion creates a wider world and an active, intense, emotional involvement in it. It creates a universe of respect for human and nonhuman beings. This allows taboos and protections to be invoked and observed.

In the contemporary world, religion is more often divisive than uniting, and in many areas and sects it has lost most of the above advantages. Our task is to find something equivalent—involving, emotionally compelling, solidarity-building—that does not depend on highly divisive dogmas and social attitudes. Many religious communities are now involved in the search; the "creation care" movement in the United States is now rapidly growing. Many nonreligious conservation movements rely heavily on beautiful photographs, solidarity-building activities, and emotional appeals—a policy seriously debated in the conservation world.

3

Rationality, Emotion, and Economics

Income disparities . . . remain enormous, with the income of the 225 richest people of the world equaling that of the poorest 2.7 billion, or 40 percent of the world's population.

(Bergman 2007:A9)

Part I: Economic Rationality in the Service of Passion

Reason is, and ought only to be the slave of the passions, and can never pretend to any other office than to serve and obey them.
David Hume
(1969/1739–40:462)

INADEQUACIES OF RATIONAL CHOICE THEORY

Most of the world's ecological problems come from the production of staple goods—things people need to survive. These are foods like wheat, maize, and rice as well as other staple commodities, from electric power to cotton. In a world in which three-quarters of the population is less than affluent and 20 percent is not even eating regularly, we have to remember that the luxury of cutting back consumption is an option open to less than a quarter of us— at most. We of that lucky fourth should, indeed, cut back—drastically. In fact, we have to; we will either do it voluntarily now or involuntarily in the near future. But, in the meantime, we have a moral and practical charge to raise the standards of living for the other 75 percent. The wants and needs of those 75 percent are also not well figured into the market economy; many of them do not take part in it more than marginally (see elsewhere in this work).

Raising the standard of living for those 75 percent will require an *increase* in total world consumption, no matter how much we of the affluent world cut back. Thus, ecological improvement must come from a really huge increase in efficiency, sustainability, and use of every scrap of traditional and nontraditional ecological knowledge that we can assemble, rather than from overall cutbacks on final consumption.

Moreover, "full market pricing" is important but will not quite do the job for sacred sites, beautiful mountains, or—most dangerous and important of all—the millions of species whose importance is unknown but whose contribution to biodiversity is so significant.

All this being said, there is a real problem with things that people consume to an irrational degree. They often demand that governments "give" them these commodities or subsidize the production of them.

We Have to Begin by Looking at the Whole Economic Question

By definition, "resources" are things we use. They are typically used to provide food, clothing, and shelter. Often they are used for more arcane reasons, such as status maintenance; the Northwest coast Indians burn valuable fish oil at potlatches, while the rich American wastes steel and chrome making stretch limos. The resource is still being used: someone is paying for it; someone is getting paid for it. Economic principles clearly explain the transactions involved. For everyday purposes, economic rationality provides reasonably adequate accounts of resource use.

The regnant theory in accounting for environmental behavior is rational choice theory (see, e.g., Field 1994). This is related to the dominance of economics in shaping modern thought. Philosophers were the "idea men"—and male they almost all were—until 1800. From Plato and Aristotle to Locke and Kant, they were the sources of the thoughts that made nations and shaped behavior. But from then on the economists and political economists took over the role. America was founded on Locke's ideas but is now run by neoclassical economic theory. This is sadly limiting. Anthropology, biology, psychology, and sociology should have some play.

The economic view, as presently defended, holds that people act to get maximum "utility" at minimum cost. "Utility" refers to anything people want: fish, fuel, books, pet parakeets, love, status, political power, and the rest. An individual is rational to the extent that she systematically seeks out the most cost-effective ways to get her preferred bundle of utilities. Ultimately, this view is totally tautological because one can always maintain that any act is done to produce some utility or other. Even the errant behavior of a schizophrenic is done for *some* reason, to accomplish *some* sort of purpose.

In practice, however, rational choice theory ("rat choice" to its many friends and enemies) is normally limited to applications in which there is a definite end in sight: a firm maximizing profits, a fisherman maximizing returns from catch, a farmer maximizing yields per input.

People get together out of sheer need for society, but they stay together partly so they can work together and exchange goods and services, thereby satisfying needs. In doing so, they quickly find that producing goods and exchanging them ad hoc is not always easy. One needs institutions: rules and principles that reduce the costs of doing business. Economists distinguish *production costs* from *transaction costs*. Production costs include the costs of land, labor, capital, information about producing things, and advertising. Transaction costs are the costs we have to pay for carrying out and recording transactions, but also for assuring trust, securing property rights, finding out if people we are dealing with are honest or not, and getting recourse if they cheat us. Economists regard institutions as set up to reduce transaction costs (North 1990). Institutions also regulate society in general and arise from negotiations of the society's members about how to live, how to be fair, and so forth. Organizations arise to implement the institutions. (Note that, in economic jargon, the term "institution" does not have its normal English sense of "college," "prison," or "asylum." Those are "organizations.")

This requires us to look at matters such as individual differences in reckoning up transaction costs. Individual people differ widely in the ways they make decisions and deal with transactions (Kroeger and Thuesen 1988; Myers 1982; Pervin 1990). Some people are more sensitive to others' opinions. Some are more emotional. Some are more angry or neurotic. These preferences are established early (and even have a genetic component) and persist through life. The child who is a rigid conformist or a go-with-the-flow original is apt to be such throughout life. Clearly, these findings pose extreme challenges to ordinary definitions of rational choice. At the very least, one person's "rational" is not another's.

Yet rational choice theory is inadequate as an explanation for human behavior, even economic behavior, as argued at length by former rational choice theorist Michael Taylor (2006). We have observed that some ecological activists have died for the cause. Others have sacrificed a great deal. Millions more have sacrificed a small amount—money to conservation organizations, cans to the recycling center, a letter to a congressperson. Individual choice theories tell us that no one will act this way. Such self-sacrifice—or even just voting—is "obviously" too costly for the tiny and uncertain benefits it usually produces.

The survival of the human species shows that these theories are wrong. Why they are wrong, however, is still not fully known. We just do not understand where people like John Muir and Rachel Carson come from. Nor, for that matter, do we have a clue about what produced Gandhi or Martin Luther King. Rational choice theory and other economic and psychological theories make no distinction between Hitler and Gandhi; both were out for their self-interest. Why were their self-interests so disparate? Rational choice theory is silent.

In spite of this, rational choice has evolved into a religion-like belief system, championed with utterly irrational fervor (see the humorous but incisive commentary by Elliott and Atkinson 2008, as well as Taylor 2006). One reason is that it can be used to justify existing relationships of inequality and especially the economic policies of throughput maximization.

Rational choice theory does not deal successfully with the question of when a person will act for a narrow or short-term self-interest as opposed to a wide or long-term one. Ecological problems very often consist of choices between immediate gratification and a much greater benefit in the future or between a small benefit at the expense of one's neighbors and a much greater benefit that must be subdivided with them. It is precisely here that we must make use of Robert Frank's theory of emotions being necessary to "drive" long-term interests and of society as developing institutions to harness those emotions (Frank 1988). Rational choice theory is more adequate for matters of short-term gain. However, even here, some behaviors can be as difficult for rational choice theorists as the problem of Gandhi's political action.

Even at its best, economic theory is an inadequate guide once we step away from such everyday matters (Anderson 1996; Rosenberg 1992). It is premised on the assumption that humans decide everything, or at least all resource-management questions, by rational assessment of alternative means to desired ends. For this to be the case, people would have to be wholly rational choosers.

If people were always rational, and given to acting in their long-term self-interest, we would never have gotten in the mess we are in now. To make perfectly rational choices (in the narrow sense), they would have to have "perfect information" and be immune to error (Elster 1983; Rosenberg 1992). Obviously, we are not. Even the experts do not always know enough to make perfectly informed choices. Most people are not experts.

Frequently, what appears to be rational choice is actually an emotional response to a perceived problem. The emotional response works—not necessarily very well—and the observer is led to believe that a careful process of intellectual calculation was involved. Culture very often mediates such responses. Culture is, among other things, a storehouse of tried and tested responses to problems. Thus, we often come to act rationally through conformity to tradition.

Clearly, rational choice theory provides only a moderately accurate guide to behavior. When it does correctly model or predict behavior, one must suspect that competition has weeded out the flagrantly foolish (Smith 1776/1910), rather than assuming that every actor has chosen wisely. One can only say, with Jon Elster, "Being irrational *and knowing it* is a big improvement over being naively and unthinkingly irrational" (Elster 1989:47, his italics). One can only, at this point, reiterate the desirability of rationality as a normative condition.

If rationality is defined as making the best decision possible on the basis of the best information available, prejudice and bias should be the ultimate irrationality. However, if rationality is defined as maximizing goals, then prejudice and bigotry can be rational. Consider a person whose most desperate need is a good self-image but who is, in fact, pretty much of a scoundrel by normal standards. The easiest and most effective way he can shore up his self-image is to think that other people are even worse and to blame them for his condition. It is thus rational for him to be a bigot and to refuse to consider any accurate information about the group he hates. (This is, in fact, a much-simplified version of one main current psychological theory of bigotry.)

Apparently, neoclassical economics was based on 19th-century theoretical physics (Nadeau 2008). The relevant theory within physics has been superseded, and in any case the application was only a metaphor. The metaphor provided the idea of the "market" as a closed system, isolated from resources, limits, and noneconomic activity, including politics—which had previously been fused with economics in the field of "political economy."

Anyone who has followed the economic advice columns in the major media is aware of the limits on the predictive power of economic theory. These limits are perhaps especially clear to those of us who have trusted the experts on matters of investment, only to learn that we would have been better off burying gold coins under the floor as French peasants used to do. (And I wrote this before 2008!) One soon realizes that economists may not be able to predict the future, but the reader can easily predict what the economists will say—on the basis of their politics. Liberal economists see conservative policies as sure ruin and liberal policies as the only hope. Conservatives, of course, predict the opposite.

Even with all these qualifications, however, rational choice theory remains a good place to start if one is modeling human choice-making. Humans have, of course, learned to live with processing limitations. First, in real-world situations, minor processing errors can be trivial or even advantageous. Gerd Gigerenzer (1991, 2007; Gigerenzer et al. 1999) points out that our mental shortcuts are valuable, even necessary, to keep us from the total paralysis of having to figure out the absolutely perfect course every time we act.

Second, people learn through experience and through negotiation. They talk, discuss, debate, bargain, and interact. Third, societies encode their learning in the set of institutions we call "culture." Often, it is our very emotionality and social conformity that lets us seem rational.

So rat choice theory remains a necessary beginning for our concern. The narrow conception of rationality as unemotional selfishness is simply wrong. On the other hand, ideas about making a thoroughly considered, studied choice of how to get a packet of utilities are useful analytically, not too far off descriptively, and also a worthwhile normative ideal to strive for. We should not expect people to make rational choices all the time, but people often approximate rationality fairly closely, and it is useful to see when and why they do this. We may even legitimately hope to help each other make more rational choices.

Determination of behavior by society or culture—the abstract entities, as opposed to actual individuals—is a favorite of social scientists. Talcott Parsons's reification of society (Mills 1959) and White's of culture (White 1949) are well known. It seems likely that these views are the actual body of theory originally opposed and attacked by Arrow, Downs, Olson, Becker, and others. The timing is right: The pioneers of rational choice were writing in the 1950s and early 1960s, when Parsonian sociology and Whitean culturology were at their height.

Sociology and culturology do not seem to have been definitively refuted by rational choice theory. We now need no extended discussion to prove that neither society nor culture can act, being social scientists' abstractions formulated from observed behaviors of individuals. Individuals belong to societies and have to respond to other individuals in those societies; individuals learn culture and make their rational choices on the basis of their cultural (and individual) knowledge base. Society and culture are not external, ideal types forcing individuals to act. Some people appear to conform mindlessly, but they appear to be rationally maximizing approval of others rather than acting without thought. The teenager who spends hours in front of the mirror in an attempt to look perfectly in style is the opposite of mindless. She is thinking terribly hard about it all. Of course, to some, she would still be "irrational." Her behavior is rational in light of her goals, but her goals are arguably narrow.

On the other hand, culture does determine a great deal that rational choice cannot undo. I did not rationally choose to speak English; I was stuck with it before I was old enough to have any choices. And similarly for a thousand other habits, from body movement to eating with fork and spoon instead of chopsticks, I might today rationally decide this was all wrong, but there is little chance of my doing much about it. Most people never even reflect on the matter.

Thomas Hobbes (1950/1651) told a fable of wild men of the forest coming together and drawing up a social contract that gave them stable government and provided peace. Hobbes saw a society of isolated individuals who eventually became weary of their state of permanent "warre" and came together to form a covenant that established rulership and ownership. They duly concluded that a monarchy was the ideal form of government and set up institutions accordingly. Hobbes gives us a fine account of the isolated individuals, whose lives, we recall, were "solitary, poore, nasty, brutish and short" (1950:104).

Of course, humans are social, not wild, men. But firms really can act like Hobbes' savages. They are wholly devoted to maximizing profit, that is, to short-term, material rationality. How do we get the world's firms and nations to sit down and agree to stop the competition that is driving them to shorter and shorter planning horizons? People have to sign up for voluntary restraint, even if it hurts their interests (by any economic calculus), or society cannot function.

A truly free market cannot exist because, in a genuinely unregulated market, the strongest man would just grab all the stuff and run. As everyone from Adam Smith on down has pointed out, even the "freest" market requires laws to secure property rights and contracts. Since these laws will, inevitably, be written by somebody and enforced on somebody else, they are not some sort of neutral, bland agent. They protect certain interests against others.

One can contrast the self-destructive behavior—massive deforestation, soil erosion, and general drawdown of resources, as well as social problems—encouraged by early capitalism (memorably described by Marx). Private ownership and the freest market in the world did not prevent the tragedy of the commons. In fact, it accelerated it. Workers' labor power was the open-access good that was destroyed. Capitalists, locked in cutthroat competition, found it not only expedient but competitively necessary to work their workers literally to death (anyone who thinks Marx exaggerated may consult John Burnett's *Plenty and Want*, 1966). Clearly, the free market does not cure the commons problem. From Mongolia's grasslands to England's (and William Blake's) "dark satanic mills," capitalism actually exacerbates it. However, socialism and communism have not only failed to check this; they have proved even worse. So far, in almost all cases, they have put the productive interests—the maximizers of short-term interest and maximal throughput—in charge of society. This led to the industrial wastelands of Eastern Europe and China and to the Aral Sea catastrophe described previously.

The Contrast with Traditional Societies Is Striking

Even with social humans, motivating prosocial behavior is not easy. How do you motivate park guards to be honest, planners to be responsible, and biologists to study ecological problems? Rational economic self-interest simply cannot do it. The transaction costs, especially for the initiators of action, are simply too great (Frank 1988; Taylor 2006).

We may wish to rehabilitate the concept of "reason" that animated Europeans in the 17th and 18th centuries. Their "rationality" was contrasted not with random action, nor with nonmaterially motivated action, but with intemperate, hateful, whimsical, or irresponsible action (Holmes 1990). It also had something to say about goals as well as means. The rational person would strive to be moderate, temperate, cheerful, just, philosophical, and appreciative of life and people. This view had its limits—enough to provoke the Romantic reaction—but it had its merits, too.

To expand the role and value of economic theory, we have to give a more adequate account of what people's "beliefs and desires" really are. We have to understand how people plan in this world of imperfect information and highly emotional controversies.

To apply economic theory to the question of saving the world's ecological base, we must do more: We must understand how people construct institutions for successful management of resources.

ADAPTATION

A slightly different derivation from economic theory is the theory of adaptation. This theory does not use individuals as the units of analysis, as rat choice theory does. Instead, it uses societies and cultures as the units of analysis and examines how they adapt.

Cultural ecologists once based their models on adaptation (Bennett 1976). However, the adaptationist paradigm has been largely abandoned in recent years. There are two main reasons for this.

First, many ecologists had a simplistic and romantic view of how harmonious, systematic, and integrated the world's ecosystems are (natural and humanized; Botkin 1990), as well as the world's societies (Robbins 2004). These ecologists saw adaptation as having already occurred, so to speak. In fact, environments are always changing, and people have to change with them.

Second, no one defined adaptation very well. In biology, adaptation means something very clear: If, in a given setting, Animal A leaves more descendents than Animal B, Animal A is better adapted. Genetic legacy is the currency. In anthropology, unfortunately, no rigorous definition became widely accepted. This fact, combined with the romantic view, led to a situation in which anything could be seen as an "adaptation."

It is possible to save the concept of adaptation by grounding it in theories of economic development. Economic development means, simply and straightforwardly, more wealth per capita. Adaptation might, then, mean more wealth per capita with less environmental cost. If a group can increase their wealth without damaging the environment, they have found a better adaptation. If they increase wealth but damage the environment more than they increase wealth, they have lost, not gained. This begs the question of how one measures "wealth" and "environmental damage." It is possible to use money, but one must then assign shadow values to poorly monetized things such as genetic diversity or enjoyment of scenery.

It would also be possible to define adaptation to include the process of learning to feed more people from the same resource base—to increase population instead of wealth, again without environmental destruction. This would have the advantage of being closer to the Darwinian definition but would have the disadvantage of making us say a culture was "well-adapted" when it was keeping millions of destitute wretches alive but in utter poverty. One is reminded of the absurd society traditionally postu-lated in philosophy classes to discredit utilitarianism: one in which a wildly happy despot rules over lots of miserable wretches, the *average* wealth and "utility" being very high! Philosophical utilitarianism check-mates this by its clause "each one to count for one, no one for more than one," but World Bank utilitarianism comes close to the philosophy class reduction.

One could generalize and say that adaptation has taken place when peo-ple use the environment more efficiently—when they get, from a given (fixed) resource base, more of anything they want, be it children, dollars, or beautiful views. This, however, takes us away from the abstract realms of the early anthropological models and forces us to attend to individual humans— their experiences and desires and their unique personalities. Adaptation thus is still a vexed word.

However, it directs attention to a real problem: how people learn to use resources. Individuals learn from their teachers and from experience. Cul-tures encode new information when it is widely known and widely accepted. What happens then? Why do people want what they want from the world ecosystem? Why do they accept some laws and rules, such as hunting limits on deer, and not others?

Anthropologists have never really believed in the static models of "unchanging, traditional" society that used to be widespread in the social sciences. Anthropology in the 19th century focused on "cultural evolution," even though they thought many "savages" had "stagnated" at a "low level." The 20th century brought an emphasis on "culture history" and "culture change" (on the recent origins of many "traditions," see Hobsbawm and Ranger 1982).

RESPONSIBILITY

All this seems to be far from the environment. It is not. Our highly emotional feelings for the environment (including aesthetic feelings, "emotion" of a sort) cannot be divorced from the emotionality of social action.

Of all the complex integrations of passion and reason, the feeling of responsibility must be the most complex. It seems wholly rational—yet it is a powerful, unbearably intense motivational state. It is not even remotely addressed by talking about guilt and shame. I feel responsible for my children not because I fear guilt or shame and not just because I love them either. I am always ready to do almost anything for them. This has nothing to do with a fear that I might be socially disparaged if I failed or even that I might hate myself. Responsibility is a function of love or concern combined with knowledge of a need for caring behavior. Yet it is also an *autonomous* feeling, *phenomenologically quite unlike any other*. It arises from interaction and from direct perception of the people and things with which we are genuinely personally involved.

In studies by Antonio and Hannah Damasio (Damasio 1994), responsibility appears to be the great casualty when the lower frontal cortex is deranged. People with such damage may know what to do but seem genuinely not to know how to feel responsible about it. The Damasios have given us a way to talk about responsibility—heretofore a distinctly underdeveloped concept in Western philosophical writing. Responsibility has been relatively ignored because it is so obviously a mix of reason and passion. To the pre-Damasian philosopher, it would have to be a "mere" consequent and derivative state. It seems like a jury-rigged combination of guilt, shame, and anticipated consequences. To the post-Damasian, responsibility may prove to be both the most basic and normal and yet also the highest of human conditions. It is the defining attribute of our humanity and of our selves.

The same goes for being interested in the world; interest is simultaneously an emotion and a cognitive cause. Both questioning and certainty are in the same immediate mental terrain: cognitive feelings.

Thus, we can base our environmental consciousness (and, of course, our human interactions too) on deeply felt interest and responsibility. A post-Damasio morality must be based on the idea that humans are, at their best, integrating passion and reason into responsibility, sociability, sensitive planning, and nuanced and complex emotional representation. They are responding to the environment both emotionally and rationally—with the full heart and the full mind or, rather, with the full and total person.

Ecofeminists have sometimes taken the contrast of male and female as biological destiny and advocated "female" input to the environmental cause in order to get emotional and "spiritual" considerations into the cause. The biological essentializing cannot stand; such sexist stereotyping is disproved by every study. But we do need to draw on the human ability to integrate both states within the human mind and to be responsible.

ALTRUISM

Much recent literature in economics and sociobiology maintains that "altruism" is rare or unknown, while "rational self-interest" (read total selfishness) is inevitable and is the human condition. The classic statements of this view are by Anthony Downs (1957) and Mancur Olson (1965). Recent sociobiological claims in this tradition include those of Richard Dawkins (1976), Matt Ridley (1996; he has since changed his mind), and countless followers. In them, we have biological essentialization of an extreme form of microeconomic theory.

It is clear from the recent sociobiology literature, and from much of the economic literature on which it is based, that there is considerable logic-chopping going on. "Altruism" is defined in an extreme way, as an act that benefits someone else but is pure cost to oneself. "Rational self-interest" is anything that benefits one at all, even at considerable cost. Thus, for the economists, doing a major favor for someone else is "selfish" if it makes me feel good. Which it does; many studies show that almost everyone enjoys it.

If we define everything the other way, the picture looks quite different. Let us take a very tight definition of rationality and say that an act is rational if and only if it has been proved to be the very best way of getting maximal benefit for minimal cost. If we do this, then only the mathematically calculated actions of very sophisticated business persons or firms are rational.

Conversely, let us define "altruism" to mean any act that benefits another but does not materially benefit the actor or the actor's immediate relatives. Doing something nice because it feels good or even because we like to bask in the resulting gratitude and approbation is still altruism. Under this loose definition, probably most of the things we do in everyday life qualify as altruistic. So definition games can be played either way.

If we compare loosely defined rationality with loosely defined altruism, we find a near-total overlap. Everything people do is done for some sort of reason. Everything people do can be claimed by some extreme sociobiologist to be good for the gene pool. Conversely, everything people do is done with one eye cocked to public opinion and the other cocked to one's own pride, guilt, and shame. All but the most extreme mentally ill behavior is rational in that broad sense. Conversely, all but the most extreme callousness is altruistic, in the sense that it is done with reference to others and how they will be affected.

On the other hand, if we take the tight definitions, then both rationality and altruism are very rare, if they occur at all. Firms maximizing profit may use their mainframe computers to make genuinely rational calculations. Humans may (and rather often do) leap impulsively into the water and drown trying to save a child or even a dog. The trouble is that we have too little idea why. Conventional theories of economics, politics, and psychology have absolutely nothing to say about these people (any more than they can explain great artists or musicians).

DISCOUNTING

Devastation of the environment is obvious to most people but still continues. Sometimes it is extremely hard to explain this except by collective insanity. We are urbanizing the world's prime farmland all over the globe. The most productive soil is the first to be covered with asphalt. Traditional cultures usually prevented this through social rules. With us, quick paper profits to developers trump the obvious needs of society over time.

This is clearly irrational in the long term, and it follows a well-known principle of human thought, demonstrated in countless psychological studies: Humans set discount rates with a high slope—that is, they value future benefits at a very high rate and discount future costs. People prefer to enjoy now and pay later, even if the enjoyment is *very* small and the future cost is *very* high (Frank 1988; the point is usefully discussed in standard works such as Field 1994 and Pierce and Turner 1991). Humans actually are less good at waiting for small rewards than are chimpanzees—the latter usually waited two minutes to get more sweets, while humans in the same experiment rarely did (Balter 2008).

For an individual, this discounting of the future may be reasonable, but for society it is suicidal. On the average, people (or at least the people studied by economists) would rather have $100 today than an absolutely firm guarantee of $103 (in constant dollars) tomorrow, but $104 would look more attractive. This is the "social rate of time preference"—the rate that society establishes as a reasonable discounting of the future (Kasting 1998: 23). Of course, the more uncertain and cloudy the future, the steeper the discount rate. In a relatively safe society, the rate is as implied by the above experiment: 3 to 4 percent. Damage a year from now is 3 percent less scary than damage now. At a constant 3 percent rate, this means that a disaster in 50 years counts only 22 percent as much as one now, and one in 400 years only 0.0005 percent as much.

Thus, people are much less concerned about the far future than the near one. Neither I nor my children nor my as-yet-unborn great-great-grandchildren will be alive in 400 years. Why should I care even 0.0005 percent? I certainly would rather have a dollar now than a promise of $100 in 10 years. The promise may be impossible to enforce in 10 years, or I may be dead, or the dollar may have devaluated to worthlessness. Anyway, I want the dollar. The future can take care of itself. Gersh (1999) emphasizes this problem and how it can work with inadequate pricing mechanisms that encourage waste.

On the individual level, typical examples are provided by alcohol and drug abuse, in which immediate pleasure drives out concern for the morrow (Ainslie 1993). The ecological equivalent is represented by deforestation. Most of the world's forests have been destroyed for exceedingly tiny returns—a crop of corn, a few head of scrub cattle, a few sales of round logs. Conservation would have led to enormously greater payoffs every year. Sometimes desperation drove people to cut trees. More often it was a wider problem of imperfect markets—subsidization of logging and log exporting, unrealistically high costs of conserving, and the like. Most often, perhaps, it was simply due to extremely overoptimistic guesses about discount rates. In any case, short-term returns were irresistible.

The more people have a stake in the future, and a degree of confidence in it, the more they will look to the long term. Thus, some sort of social justice is necessary in ecological planning.

Another form of discounting is ignoring interests that involve very little money even now. In environmental economics, Mohammed Dore and Timothy Mount (1999) stress the equity issues, including the way in which the interests of the poor are automatically undervalued because damage to their livelihood and income is so economically trivial. A loss to Bill Gates of $2 billion—a tiny fraction of his annual income—is twice as large on the economic spreadsheet as the loss of $1 billion dollars to the economy of the African Sahel, where that sum represents the total combined income of about 4 million people. (Fortunately, Gates has been duly concerned, and is donating accordingly.)

These various reasons for discounting the future, and even discounting the wider effects ("externalities") in the present, make ecocide look profitable. The "jobs-versus-environment" argument is usually based on such arguments (see Goodstein 1999 for a thorough, conclusive discussion). Obviously, if the environment is really being trashed, people and their jobs will suffer accordingly in the long run. The ultimate damages far outweigh the short-run benefits. It would seem that only a madman could believe the jobs-versus-environment line when it justifies wiping out for paper or disposable chopsticks a forest that could be logged sustainably for valuable woodworking and construction material for millennia or when it justifies fishing to extinction a fishery that could be profitable indefinitely if properly managed for sustainable harvest. Yet sober people make these arguments every day.

In fact, working biodiversity and other conservation goals into economic development projects does not affect their success. Peter Kareiva and collaborators (2008) found that World Bank projects that took conservation into account were as successful as those that did not. This is based on evaluation of over 11,000 projects over 60 years. The World Bank may not be the best developer going, but it is the biggest, and it has often succumbed to the jobs-versus-environment argument in the past.

Clumsy and misguided attempts to protect environmental resources have indeed done economic damage without bringing returns (West, Igoe, and Brockington 2006), but, at least in my experience, these were cases of "jobs versus stupid politicians or stupid environmentalists," not jobs versus environment. The obvious solution is reasonable cost-benefit accounting, with the externalities taken into account. This would imply that we have perfect information. In the real world, perfect information is never available, and anything remotely like it is often difficult to find. Thus the prudent manager will be highly conservative about risk-taking.

IMPERFECT MARKETS AND EXTERNALITIES

Markets for environmental services are generally imperfect. Thomas Knoke et al. (2008) point out that tropical forests are worth over $2,000 per hectare per year in environment services (carbon absorption, watershed protection, timber, and so on) but are being cut down for "often less than US $100 per hectare" in benefits. The problem is that local farmers, loggers, and ranchers get the $100 but have no way to tap into the $2,000. In the unlikely event that anyone is paying for the ecosystem services, the money is fairly sure to go to the national, or at best local, government. The farmers on the ground see none of it. Knoke et al. go on to point out that preservation even with compensation would be inadequate since the farmers would not want to "stand by and twiddle their thumbs."

This sort of failure to pay benefits is one side of the imperfect-market problem. The other is failure to internalize costs.

Big operators displace their costs of production onto weaker bystanders. The rich owners live upwind of their polluting factory; the poor live downwind of it (Murphy 1967). As the phrase goes, "the poor are down—downstream, downwind, downtown." This displacement of costs is known as "externalization" because the costs become external to the firm doing the polluting. The costs become "externalities" from the standpoint of the polluter (Murphy 1967).

Also, a great deal of damage is done by people who operate on such a small scale that one person's actions make almost no visible difference. In this overpopulated world, millions of people, each taking only one fish or one flower, can soon wipe out a resource. Here the specification of costs and the moral issues can be fiendishly difficult. The tiny and morally almost unassailable "take" of fish and whales by native peoples under treaty rights, for instance, can be used as a "slippery slope" argument by both advocates of overfishing and advocates of total bans.

In fact, however, much or most of the environmental problems of the world are due to the ability of powerful interests to reap the *benefits* of destructive practices while passing on the *costs* to the poor.

The prototypic example is the upstream/downstream problem (Murphy 1967). The simplest model is of a farmer dumping sewage into a river. This farmer saves the costs of dealing with sewage, but farmers downstream must deal with the polluted water. Typically, the total cost to the system is greater than the cost of cleanup would have been to the original farmer. Sometimes the total cost is *very* much greater—for instance, the pollution may start a disease epidemic that kills half the downstream population.

Upstream users dump garbage into the stream; it saves the costs of cleanup and pickup. Downstream users are stuck with the costs. Unless the downstream user is powerful enough to sanction the upstream users, there is not much to be done. Riverside, California, once dumped its sewage into the Santa Ana River. Orange County—which is just downstream—forced, through legal action, progressively more and more sewage treatment. Fortunate are those downstream communities that are, like Orange County, richer and more politically powerful than their upstream neighbors.

Air pollution is similar: A factory pollutes the air; many suffer. In the nature of things, property values upwind become greater than the values downwind. The rich then live upwind, the poor downwind—as anyone can observe in Los Angeles (see, e.g., Pastor 2001, Pastor et al. 2002), whose residential patterns were produced to a great extent by the smog of the 1950s and 1960s. The rich then have every reason to fight taxes for pollution control. The poor have every reason to support such taxes but are politically weak.

Consumers do the same thing by throwing their garbage on the public road. Someone else has to clean it up. The damage done by having garbage everywhere is typically greater than the costs of cleanup to the consumer. Again, extreme cases occur when health is an issue. Discarded syringes are often reused by drug addicts, for instance, and this spreads disease. So, to save a few cents' worth of effort cleaning up, people throw away syringes that kill or ruin other people. (This is rare in the United States, but common in the Third World [personal observation].)

Another problem is that of who pays the price for ecological damage. Kendall Thu (1998), among others, finds that farm workers are notoriously denied any political voice—many of them are not even citizens of the nations where they work—and there is little hope that their concerns will ever be addressed. The situation is bad enough in the rich nations but far worse in Third World countries, where farm workers routinely handle deadly insecticides with no protective equipment at all—as I have often observed in Mexico (cf., Wright 1990). They have no way of knowing better; no one tells them, and even if they are somewhat literate, they cannot understand the technical language of the warning labels. Moreover, they are often working for large firms that leave them no choice in the matter.

In a market backed up by fair laws to protect private property, the weak can (theoretically) get together, economically and politically, to make producers internalize the costs of production. Resource prices would reflect real costs. Unfortunately, this never seems to happen, even in (relatively) free-market polities like the Hong Kong of old (where I conducted research on the matter). The powerful can manipulate the legal system and the market system to pass the real costs on to the less powerful. The system then tends to become more and more stacked against those who are already less able to exert power. Thus, in all real-world systems, the powerful—whether they are giant feudal lords, capitalist firms, or communist cadres—can distort the system, such that their activities pay well, typically at the expense (environmental and otherwise) of everyone else. Their real costs are passed down as externalities.

The most successful market economies have had the sense to see that they have to guarantee human rights in general, *outside* the market. The same is now true of the environment; we have to guarantee a livable environment, *outside* the market, or all economies will self-destruct.

YET ANOTHER REASON: THE TRAGEDY OF THE COMMONS AND ITS KIN

The prototypical "rational choice" problem in resource management economics is the Tragedy of the Commons, made famous by Garrett Hardin (1968; cf. Field 1994, and the particularly good discussion of what this model will and will not do in Borgerhoff-Mulder and Coppolillo 2005). Hardin originally used, as his model, the English common pastures. He postulated that grazers would stock more and more animals on the land until it was ruined by overgrazing. Each herder gained by adding an animal, but no one benefited from withdrawing his or her animals, because someone else simply added more. Thus, in the Tragedy of the Commons scenario, a group of users devastate an open-access resource, even when they know they are doing so, because they have an incentive to continue using the resource and no incentive to stop. This condition typically holds in fisheries (Anderson 1996; McEvoy 1986; McGinn 1998; Safina 1998). If one person, or ten people or a hundred, become too moral to keep fishing, the other fishermen simply expand their catching effort. New entrants may even be lured into the fishery, expecting that more slots are now open for them. Most of the world's fisheries have faced this problem. World fish catches, indeed, are now declining for precisely this reason.

In some fisheries, the number of fishermen can be reduced more than 99 percent with no benefit to the fish because the few remaining fishermen can use their gains (from lack of competition) to upgrade their boats, nets, and other equipment. Something very close to this happened in the British Columbia herring fishery, for example. If regulating agencies set limits on the time spent fishing, this merely provides another incentive to upgrade. We are now confronted with the depressing spectacle on the British Columbia coast of herring fishermen with million-dollar boats that may be allowed to operate for only *40 minutes* out of the year (Anderson, field research; J. Poznikov, personal communication). The herring continue to decline drastically, and the fishery will soon be a memory. An excellent book on this matter, *The Fisherman's Frontier* by David Arnold (2008), puts this in full historical perspective.

It is always necessary to point out that Hardin is *not* postulating "greed" or "competitive individualism" or any other psychological factor here. If the fishermen are—like most fishermen—interested only in feeding their families, they will act the same. If they are fishing solely to get fish to donate to the starving poor or the Sierra Club, they will *still* act the same. The logic of an exhaustible, open-access resource forces them to act this way. Hardin explicitly rejects psychologizing assumptions (Hardin 1968, and also personal communication, 1992, in response to my directly questioning him on the point). There are still those who believe that the Tragedy of the Commons is a "capitalist" affair that can happen only when greedy capitalists get together (Merchant 1992). This is not the case. Hardin's model does not require this, and, unfortunately, his model predicts all too well the behavior of unregulated fishermen in communist states and in even the most conservation-conscious traditional societies.

Interestingly enough, this situation did *not* characterize the English commons or most other grazing commons either (McCabe 1990, 2003, 2004; McCabe et al. 1992; McCay and Acheson 1987; Netting 1981). When overgrazing does get out of hand, it is often in pioneer situations with notable lack of environmental justice (a classic case is that of grazing in early colonial

Mexico; Melville 1997). Most common pasturers work out some kind of specification system. This will be discussed in the following chapters. However, tragedies of the grazing commons do occur; there have been many cases on public lands in the United States, especially in the open-range days of the 19th century.

Usually, people tend to regulate their fishing, grazing, hunting, or other offtake activities (McCay and Acheson 1987; Ostrom 1990, 1999). Hardin came to realize this (Hardin 1991). When people fail to regulate their take, one suspects active government interference in their behavior (Anderson 1987). Most of the cases of environmental injustice are due to the opposite of the Tragedy of the Commons; they occur because powerful interests exert top-down control.

Hardin's model provides an example of an *emergent:* a social phenomenon that is not predictable from the behavior, or in this case from the intentions, of the individuals in the society. Hardin originally recommended privatization. A private owner would, he believed, take good care of the resource base because the problem of open access would be eliminated. This prediction has turned out to be incorrect. Private ownership does not necessarily stop the Tragedy of the Commons.

In some cases, private owners control too small a share of the resource, in which case the Tragedy of the Commons goes right on. This occurs, for instance, with soil erosion. If a number of small ranchers allow overgrazing on a watershed, gullying and headward erosion soon destroy the good pasturage. No one rancher has enough land to do much prevention work on his own. If he tries, erosion gullies spread from neighboring fields and soon wipe out his efforts. This problem is conceptually exactly the same as a Tragedy of the Commons. Yet there is no open access and no commonly owned good. The Soil Conservation Service succeeded in dealing with this in the 1940s by dialogue between public advisors and private owners. It got them to work together. Later it came to use, locally, more heavy-handed tactics, and resistance surfaced.

In other cases, owners may not have long-term goals. A revealing case centered around the acquisition of the Pacific Lumber Company, a sustained-yield harvester, by Maxxam Corporation, a fly-by-night corporate-raider firm. Maxxam set out to clear cut the Pacific Lumber holdings, in a way that would have deforested the land for decades (if not permanently), to pay off debts incurred elsewhere. Their idea was simply to treat the trees as a wasting asset—to draw down the stock and abandon the land (Rauber 1994). Here, one private firm acted like Hardin's ideal private owner; the other acted like an unregulated user of an open-access resource. The latter firm won out, through Machiavellian behavior in an imperfect market situation. The State of California eventually stepped in, bought the land for a high price, and made it a reserve. The taxpayers thus wound up losing both the timber resource and the huge amount of money spent buying the land. Reasonable economic policy would have left Pacific Lumber Company in control in the first place, and we would still be reaping the benefits of sustainable forestry, which had been preserving wildlife, watershed, and timber values.

In still other cases, a resource may be beyond any conceivable privatization. Migratory birds provide an example. Significantly, these were among the very first biotic resources to be successfully conserved. Management started with the Migratory Birds Protection Act and is continuing today. International treaties (Barrett 2003) have served to save most species, even those on the verge of extinction. Today, the massive threats of pollution and habitat destruction may overwhelm all efforts, but at least the last 90 years have shown that these most uncontrollable and unprivatizable of all resources could be successfully conserved over a long period of time.

Even the question of privatizing the grazing commons is far less clear-cut than Hardin thought. Privatizing the commons in Inner Mongolia has been an unqualified disaster. Excellent studies by Dee Mack Williams (1996a, 1996b, 2000) have shown that more and more specification of the commons led to more and more overgrazing and erosion. Ancient commons arrangements did have problems of commons abuse; Hardin was not entirely wrong. But there were traditional ways to cope with these. Communism under the modern Chinese regime led to more commons abuses. In an attempt to stop this, the communist government introduced enclosure and privatization in the late 1970s. Individuals tended to bag off especially good plots, develop them, and graze animals on the rest. This allowed overexpansion of herds and of grazing on vulnerable sites. In 1988, new and more comprehensive reforms were introduced to stop these abuses. The new reforms provided for more specification. They made things worse.

Lack of capital has been a recurrent problem. Without money to provide water, soil erosion control, really wide-scale fencing, and the like, the pastures cannot be privatized as successfully as American pastures are. Turning this argument on its head, one can see traditional communal regulation as an adaptation to a situation in which the land-to-capital ratio is exceedingly high.

But having lots of capital, full privatization, and full control of one's fief is no cure. American ranchers often overgraze their land seriously. This is largely a result of the strange subsidy structure of American ranching. Research by my students Kimberly Hedrick and Monica Argandoña has shown that large corporate ranchers and hobby ranchers do most of the overgrazing these days. They reap the subsidies and the tax breaks. Ranchers who actually have to make a living by ranching cannot afford to let their range decline (Hedrick 2007).

We may turn to fisheries for more illumination (Pinkerton and Weinstein 1995). James Acheson's work on the Maine lobster fishery is one of the best, most long continued, and rigorously detailed of many studies of local and informal fisheries management (Acheson 1998, 2006). The Maine lobstermen regulate their take voluntarily and police their fishery informally but effectively.

If I may invent a couple of terms, I may label two close relatives of the Tragedy of the Commons: the Downward Ratchet and the Moral Cascade. A Downward Ratchet occurs when people can shorten their vision and get away with it. They compete in progressively shortening and narrowing their planning horizons. It is a Tragedy of the Time Commons. This replaced the Tragedy of the Commons in world fisheries management when nations took control of their seas; they simply discounted the future and ran down the fisheries, leading to even more rapid collapse than the open-access schemes had done (cf., figures in Worm et al. 2006).

Firms competing for market share do the same; any firm that sacrifices immediate advantage for long-term benefits is outcompeted in the short term and driven out of business. Democratic governments must dispense favors *now* to avoid being voted out of office; hence the universal but economically irrational rounds of tax cutting and pork dispensing that please voters today but guarantee disaster tomorrow. Even tourism is a problem. The park with the most tourists gets the money or appropriations, so parks are often forced to suffer far more traffic than they can accommodate, as visitors to Yellowstone or Yosemite can attest.

The Moral Cascade, like the Tragedy of the Commons, is a real-world instance of a Prisoner's Dilemma game. In a Moral Cascade, a moral institution is so impacted by cheaters that it breaks down, and morality stabilizes at a "lower" (more individual, less socially beneficial) level. The fall of the Soviet Union led to the loss of controls on Caspian Sea fishing. A scramble for fish, especially sturgeons, ensued. Many fishermen and local governments tried to intervene but were overwhelmed by the progressive breakdown of social institutions and individual consciences. The sturgeon fishery was soon wiped out. It was a Tragedy of the Commons—with the difference that the commons had been well regulated for several generations. The regulations were still there; they were just not enforced.

The Tragedy of the Commons can be seen as a transform of the upstream/downstream problem. The latter case involves transferring the externalities downstream geographically. In a Tragedy of the Commons, the externalities are displaced temporally. Future users pay the real costs of production. Today's users get the benefits.

At the very least, even if no one free-rides or takes advantage of a commons, society has to decide who shall do what tasks, endure what risks, and face what shortages (Elster 1993:55; 1994). At this point economic theory genuinely breaks down. At best, it gives us the gloomy message that the poor, being least able to resist, shall bear all the costs. Also, economists are simply not able to set a proper discount rate for centuries in advance or to calculate the costs of exterminating species that may be the only cures for yet-unknown diseases.

It is at these points that morality, emotion, and responsibility must enter.

LONG-TERM AND WIDE VERSUS SHORT-TERM AND NARROW

These are examples of a more general problem: the long-term and wide (LTW) versus short-term and narrow (STN) tradeoff (Anderson 1996). Usually, the long-term and wide interests are ultimately much more numerous and significant than the short-term, narrow ones, but a direct, immediate, "point" interest attracts attention. People with an immediate vested interest in cutting an irreplaceable forest or catching the last few cod will have a direct stake. The public, which stands to lose enormously over time from losing a valuable resource, has little direct immediate interest. No one person is threatened enough to stop the damage.

This is where emotion, morality, and responsibility come in. Unless they motivate *solidarity* among the public to outweigh the point interest, the point interest always wins (Murphy 1967).

People also, obviously, have to care. More than that, they have to see themselves as part of nature. There was once no alternative; people had to work with nature to produce food. Conversely, preservation as an ideal did not exist until the romantic movement of the 19th century. The separation of people from nature has a long history, going back to ancient times, but it was not a reality for ordinary farmers and craftspersons until more recently.

A particularly frightening example is presented by the worldwide conversion of prime farmland to housing and shopping centers. The effects of this, in a world of rapidly growing population and tightening food supply, are obvious, but in most cases no one can do anything about it. The immediate interests of the farmers selling their land and the developers buying it always outweigh public concern. Land has been saved in a very few cases in which it is truly desperately limited, such as the Netherlands' tulip fields. But even the controls on developing the farmland of the lower Fraser River valley in British Columbia—the only good farmland with a nearly year-round growing season in all of western Canada—have effectively collapsed. Housing is rapidly eliminating this resource (as I know from personal observation over several decades).

This vicious logic makes throughput and waste more profitable than efficiency. If one can pass on the real costs of production as externalities, one makes more profits and can beat out the competition. Thus, unless there is a powerful public reaction, firms get into a competition to see who can pass on the most externalities to the public. Especially deadly is the situation in countries like China, where the government runs the factories or is in league with them and actively supports industry against the public (Economy 2005; Smil 1984). Then pollution is truly out of control, and mass deaths result.

PERVERSE WANTS

Emotion and irrationality have their own costs. Quite apart from the problems of individual rationality gone out of control, there are plenty of cases in which individual wants are directly counter to all economic sense.

One insane chapter in the sorry history of human mismanagement of resources was the beaver rush of the 1820–1840 period. Beaver fur was used to make felt hats until shifting fads popularized the silk hat. Millions of beaver were trapped all over North America, with little thought of conservation—except among the Native Americans who, along with the beaver, were the immediate sufferers. (To its credit, the Hudson's Bay Company acted more responsibly, saving beaver in some areas. See Bryce 1968 [1904].) The result was not only destruction of the beaver populations but literally billions of dollars' worth of damage to North America's drainage systems (see, e.g., Rea 1983; Wohl 2005). In Arizona, streams turned into gullies; millions of tons of soil washed away (Rea 1983). Thousands of square miles turned from lush fields to desert. Moreover, beaver dams had been essential to the ducks, fish, meadow grasses, and other water-loving organisms of the continent, and they were decimated. All this so that a few rich people could wear funny hats.

Beavers have been reintroduced in some places. Unfortunately, in the interim, people had built houses and businesses in land that should by rights be flooded by beaver ponds. Beaver reintroduction has halted accordingly. Here, as in so many other cases, the results of behavior are irreversible.

It is a human desire—hypertrophied in modern commercial society—to express nonmaterial things in material ways. Much of the world's consumption is wasteful, from triple-wrapped presents to overlarge restaurant portions that nobody ever finishes. Not only is the world using 44 percent of its grain to feed animals (State of the World 1999; Evans 1998; Pollan 2006), it is not even using much of the animals. Americans in particular have turned against organ meats. Most of the carcass of a butchered animal winds up in dog and cat food.

Status consumption fuels much of the waste in the United States. "McMansions" are highly visible and are intended to be. At other times, the damage is less obvious but more serious.

Particularly pernicious is the situation in which a status good becomes a necessity because most people eventually have one; then the public goods it replaces are run down and abandoned. The automobile is the classic case. It moved from luxury to convenience to necessity. Public transport was abandoned in old communities and never begun in new ones. The car is now a necessity, not a luxury, for North American workers and families unless they live in one of the few cities that still have adequate public transportation. Elsewhere, even the possibility of bringing in public transportation has now been ruled out because zoning based on the assumption that everyone has a car has ensured that workplaces, grocery stores, schools, medical facilities, churches, and temples are all widely separated from each other. A normal daily round of errands is impossible without a speedy vehicle. The rest of the world is rapidly moving in the same general direction.

This is, of course, a fact with a history. Oil and automobile companies fought long and hard for subsidies and against public transportation. The public, always desirous of cars and typically wanting the fancy new models, went along.

The same thing is happening now with personal computers and cell phones. They went from luxury to near-necessity in a couple of decades. Yet these machines, from production to the breaking-up centers in China, where life expectancy is only a few years (see, e.g., Shell 2008), have serious environmental costs.

Hence, even the impoverished and the conservation-minded are forced to buy and own more and more things simply to function at all. One cannot reasonably attack the "consumer society" as if people had full and free choice in the matter. One can better direct attacks against politicians who continue to favor and subsidize a few high-throughput crops and industries at the expense of the public. Diversity, efficiency, and low-throughput production are not so much matters of individual choice as matters of public choice.

Of course, any want, at least any want more technologically demanding than singing to one's homemade banjo, is potentially destructive. If everyone were to go hiking in the park, the park would be dust in short order. But some wants and forced needs are much more destructive than others, especially when the trivial status they give is compared with the huge damage they do. Status consumption forced up the price of caviar, leading to extermination of sturgeons. In the 1990s, Paul Prudhomme's recipe for blackened redfish supposedly became so popular that it almost exterminated the Gulf of Mexico's redfish.

Another source of trouble is one of the most loving and cherishing of human activities: giving gifts. Christmas presents an orgy of material consumption not because people are greedy for material objects. Everyone knows that many—perhaps most—Christmas presents go right back to the store, into the garage, or for future "regifting." But people have to give and receive them because we use material goods to instantiate social ties.

In an ecologically sane world, no one would do this; the gifts would be songs and tears, flowers and plants, love and hope—most certainly not the trash that we all know and fear. In a world where real costs and benefits really mattered, there would be singing instead of buying records; sharing love and knowledge instead of exchanging fossilized fruit cakes; walking for pleasure instead of driving; cooking from scratch on a simple stove instead of the proverbial Yuppie pattern of "having a kitchen full of gadgets and eating out every night."

Marcel Mauss (1990 [1925]) pointed out the complex nature of gift exchange many years ago—the many social messages that go with gifts. He pointed out that they not only express social solidarity; they demand a return, in goods, services, and personal loyalty or bonding. This is all to the good, on the whole, but not when it leads to the wholly gratuitous consumption of millions of tons of material.

Any shift from expressing emotions or finding security via material things to finding them in reality (i.e., as intangible and immaterial qualities) would be an improvement. And it may not be out of place to suggest that gifts consisting of donations to conservation organizations would be infinitely more helpful to the world than gifts of crystal bowls and electric pineapple corers. I am impressed by the number of memorials, in particular, that request gifts to scholarship funds in lieu of flowers.

There are many reasons why it is fortunate that the United States is the biggest world power, but one less fortunate side stands out here: American culture is, and long has been, noted for some degree of materialism and conformity. Some of this stems from a puritanical religious heritage, some from the pressures of individualism. Individualism is stressful; people need to be socially grounded, and the more insecure ones conform abjectly to get that grounding (Riesman et al. 1953). We consume a great deal; we also consume narrowly. We eat only a small percentage of animal and plant species, and a uniquely narrow percentage of the parts of the species we do eat. We insist on watching the same few movies, the same few TV shows. We drive similar cars. We wear plain, dull, identical clothes. We appear to prefer "virtual reality" to nature or to art.

American consumption patterns have spread rapidly. Fast food as an institution is popular worldwide, largely because of its associations with America—the center of power and material wealth (Watson 1998; see also Rifkin 1992). People throughout the world eat just as they watch American movies, wear American-style clothes, and practice American English. This is basically status emulation, but it is sometimes more than that. Many people have a literally magical belief that imitating Americans would make the imitators more successful and wealthy (Watson 1998). I remember being told quite directly in Malaysia, when I asked why people wore hot Western clothes in a tropical climate, that "we do it so we can be rich like the Westerners."

Sugar (Mintz 1985) presents a similar case from earlier times. Desire for sweets began as human nature. It rose with the high status of sugar as a luxury, then peaked when the price of sugar rapidly fell, making it now one of the cheapest sources of calories. The world duly emulates the rich West and switches from fruit to cookies and soft drinks. Another line I heard in Malaysia, when I asked about westernization, captures the reality: "Candy and cookies are made in factories, but fruit comes from *peasant villages!*" That last phrase was spoken with disgust and contempt. A combination of status and cheapness has made soft drinks a leading calorie source in much of the world today.

Other perverse wants are more general, such as the continually expanding demand for paper that is leveling more and more of the world's forests. This could be solved: We could just make all that pressed-out sugar cane into paper. Electronics promised the "paper-free office" but wound up merely making it easier for paperwork-loving bureaucracies to demand more and more.

All this may sound puritanical, but in fact Puritanism is the source of much of the problem. It teaches that one should suffer to get salvation and that salvation is shown by wealth (Weber 2002 [1907]). Those elderly people faced with the care and maintenance of a huge, ugly estate, those tropical businessmen in dark wool suits, and those listeners of modern pop music are all suffering to show their blessedness—their status as consumers. No one enjoys the excesses of modern material civilization. If people chucked it and did what they enjoyed, life would be simple, music would often be homemade, and the environment would be carefully preserved.

UNNATURALISM AND THE LAWN

A fine case in point is the lawn. It not only requires resources, it also demands countless hours of maintenance, usually put in by miserable suburbanites seeking virtue through conforming to community standards.

A third of all the "agricultural" spending in southern California is on ornamental gardening, mainly lawn-keeping. Lawns use millions of tons of pesticides, destroy natural habitat, and are in every way an ecological disaster. In the United States, they cover more acreage than any food crop (Bormann et al. 1993; Jenkins 1994; Uhl 1998:1175).

Grave authors who have considered the matter recommend replacing the grass with natural plants, vegetable gardens, roses, fruit trees, bushes for bird and other wildlife habitat, even concrete—in short, anything. But every American has to have one. (I admit to bias in this situation: I am violently allergic to grass pollen. I replaced my lawn with native plants and with roses, but then I moved—and now I have to start all over again.) Lawns cover an area of the United States the size of Pennsylvania and are our largest "crop" (Jenkins 1994). They are dosed with disproportionately large amounts of dangerous pesticides, many of which escape to the wider environment. Since perfection, rather than economic success, is the goal, levels of pesticide are often 10 or more times what they are on intensively managed farms.

One town—Seaside, Florida, which is built on sand dunes that will not support grass—has banned lawns and enforced native plant landscaping (Pollan 1999). Desert-style landscaping is gaining ground in Tucson and Phoenix, where water shortage has forced "xeriscaping" on the public (once again, my personal observation). By contrast, a vast number of American communities enforce rigid standards of lawn and yard maintenance.

This is part of a wide movement. One odd bit of "revealed preferences" is the striking rise of unnaturalism in the late 20th-century and early 21st-century Western world. In the 1960s and 1970s, "the natural" was idealized. Some of the "natural" was rather far from anything found in the wilderness, but at least the idea was there.

After that, a striking counter-reaction took over. Fashions ran to dyed hair, then to body-piercing and tattooing, and finally to "extreme makeovers" and runaway plastic surgery. Hiking gave way to off-road motor vehicles; at least in California, the latter now make up the vast majority of recreational traffic on trails where they are allowed—at least on the trails I (try to) hike. Hollywood movies, more and more contrived, take over from homemade entertainments like storytelling and making music. Video games and animation replace actual sports and games. Strangest of all is the boom in Caesarian sections, which have about 25 times the risk of death of natural birth and which cause major permanent scarring and damage (Wagner 2006). Doctors want them because of time and money issues, but women themselves want them, thinking that anything unnatural and technological must be safer, easy, quick, painless, or less damaging to the figure.

The unnaturalist tradition in Western civilization is, however, much older. It was established by the Greeks and above all by the Romans, who idealized city living and hated and feared the wilderness. Modern nature-friendly views, in the West, are derived from the Celtic and Germanic north of Europe, as well as from Asian influence. The "civilized" Roman distaste for nature has tended to prevail through European history. Its current boom is clearly more related to modern advertising and corporate profits than to Roman values, but without the basic philosophic grounding the corporations would not have had much purchase.

CATS

Sometimes animal lovers fight each other, and the result can be a worst-case situation for all parties. This has occurred in the case of feral animals. Strong lobbies protect feral dogs, cats, goats, burros, and horses, in spite of environmental damage they do. A more telling case is cats. There are at least 66 million pet cats in the United States, as well as perhaps 40–60 million feral ones (Gay 1999:5). Cats kill huge numbers of mammals and birds; in Wisconsin they kill somewhere between 20 and 150 million birds a year and many more mammals. Cats in Tucson kill about 80 animals a year per cat (Gay 1999:6). *Science* in 2007 reported that America's cats kill an estimated 1 billion birds a year (Science 2007).

Many of the kills are pests like house mice and house sparrows, but many are not. I have observed that all ground-dwelling birds are eliminated from areas where cats are kept in high densities. I did a rather informal study of birds around the married students' housing at the University of British Columbia, which is particularly cat-rich, and found that the number of ground-nesting and low-bush-nesting birds (individuals and species) decreased steadily as one approached the housing area, dropping to zero within a block of it; birds that nested high in trees did not show this effect.

Worse has been the problem of lizards. In my area of southern California, the local Coast Horned Lizard and Orange-throated Whiptail were once abundant but are now endangered. Cats are a major cause of their decline (personal observation—a great deal of it over 50 years). The fact is that some unrestricted cats are very savage hunters, even when well-fed. This being noted, the National Audubon Society has released a resolution to urge people to keep their cats indoors, spay and neuter them, and control feral cats. Other groups have acted similarly, and some states, including California, are doing likewise (Gay 1999).

However, in talking to cat owners, one finds that many of them take the attitude that it is "natural" for cats to hunt. Confining them (even the diseased feral ones) would be cruel. Even at a "Religion and Animals" conference I attended in 1999, a conference devoted to animal rights, I found that no one wanted to touch the issue of cats. Thus, one group of animal lovers (cat lovers) are set against another (bird and lizard lovers). Solidarity is damaged yet again.

SPECIFYING COSTS

The obvious conclusion of all the above is that, whatever else we do, it has to begin and end with specifying the costs of a process on those who get the benefits. Polluting firms must clean up their pollution. Cat owners should pay for bird protection and restoration. It is simply not fair for them to go unscathed when plants die for lack of birds to eat insect pests. We can no longer afford to take a narrow view of politics, in which special interests get what they want and the diffuse public interest can be ignored.

One result of taking a narrow view of "politics" is that most political-ecological studies read like morality plays: an idyllic local community, presumably free from politics, is increasingly subjected to the evil machinations the wider world. Liberal writers tend to take this world to consist largely of Wall Street, the CIA, or the capitalists. Conservative writers see "big government" and "communism" as the villains. Either way, powerful, evil outside forces are blamed for everything.

One trouble with morality plays is that they have no place for mistakes or for human foibles like status consumption. In the real world, foolish mistakes by well-meaning people can cause havoc—especially in environmental matters.

Many people see "indigenous" and "traditional" people as "different from us." All too often, they are portrayed as living in an idyllic prepolitical space. Alternatively, others see them as foolish wasters who kill game wantonly or who cut down a tree to get one fruit (a classic stereotype, but it does actually happen on occasion—see Headland 1997). In fact, traditional people are people and act like people. Stereotypes dehumanize and distance them. By trapping them in an ideal but doomed world, utterly different from "ours," stereotypes write off the common humanity of their experience and the possibility that their knowledge can benefit us today (Kay and Simmons 2002; Wolf 1982).

The worst problem with the simplistic scenario is that the environment is itself a player and so are local history and other incident factors (Vayda and Walters 1999). A world in which cropland has shrunk from half a hectare per person in 1960 to a quarter-hectare today (Pimentel 1999) is gong to have land conflicts. A world in which 40 percent of crop production is still lost to pests (Pimentel 1999), as was the case before modern pesticides entered the picture, is going to have food problems, politics or no. Pesticides have allowed us to more productive but delicate crops. So yields are raised but losses remain high. Economic realities then dictate that the poor starve. "History"—that is, causal factors that do not operate today but once operated in the past—is responsible for things like crop choices. Politics merely distributes the suffering more unevenly; the powerful avoid it, while the politically weak must endure far more than their share of it (cf., Mokhiber and Weissman 1999).

Political Ecology

GIANT FIRMS

The record of traditional societies is that they always find some way of managing for the long term (if only by not managing much of anything). They find ways of uniting to get management goals accomplished.

The universality of this among all earlier societies shows that something is definitely strange about the present. We have vastly more economic and scientific resources, yet cannot keep ourselves from polluting, wasting, and overusing. Traditional societies were imperfect, but they tried.

Individual choice, as in the cases of beaver hats and lawns, certainly bears a major share of the blame. But, also, much of the problem is the sheer political power of the giant primary-production interests that live by short-term managing. These are the winners in past Downward Ratchets and Moral Cascades. They can be governments, small or large firms, or giant, government-backed, multinational interests.

In the United States and many other countries, a disproportionate share of environmental damage is done directly or indirectly by a very few operators, all highly subsidized by the taxpayer (Myers 1998; Zepezauer and Naiman 1996). Giant logging, mining, grazing, agribusiness, real estate development, and energy firms are overrepresented among creators of environmental problems worldwide.

This is *not* to say that all giant firms are bad, still less that all small ones are good. Obviously, many big firms are very good corporate citizens, and many small firms use their modest size to "go under the radar" and get away with a great deal. Moreover, the firms' productive activities are often necessary; people have to eat and stay warm. And the less necessary activities are often driven by consumer demand, so one can always blame the consumers. Indeed, the consumers do bear their share of the blame. And the giant firms often do a good job of providing genuinely necessary goods and services.

However, large, subsidized firms tend to be inefficient, are often above the law (at least the environmental protection laws), and can use advertising, political patronage, and outright corruption to manipulate demand. They can, for instance, persuade governments to subsidize them and then to buy their products at inflated prices. In today's world of truly giant multinational corporations, one huge firm can do a great deal more damage than dozens of small operators. Also, subsidies and tax structures, or outright government ownership, make the biggest players less accountable to either the market-place or government regulation. As long as citizens do not unite to demand accountability and efficiency, this will continue to worsen.

The problem is systemic. One problem with large productive systems—communist governments or giant firms—is that it is easy for bureaucrats to take over decision making. Bureaucrats (as defined here) are administrators who are not directly accountable for their acts. They neither have to make profits nor do they usually suffer for their mistakes.

Giant firms also benefit from the complex and top-down structure of the world economy. They are often literally above the law; they cannot be regu-lated by any one nation, since they can simply move money and offices around the world. They can, too often, bribe governments. They can spread costs thinly over a vast public. They often find it profitable to destroy one resource totally to get money to invest in something else. Thus, timber com-panies often destructively log out forests and reinvest the resulting profits in insurance or real estate, a bit of ecocide that has sometimes been made attrac-tive by poorly designed environmental protection laws.

The food crisis of 2008, for instance, enormously benefited the giant multinational agribusiness corporations, almost doubling their profits. For the statistics, and also for very different takes on the situation, see *Grain* (2008) or *Wall Street Journal* (April 30, 2008:1). The multinationals not only were well placed to take advantage of it, they had known it was coming and had stockpiled grain while they and others speculated and traded in futures.

However, the crisis was itself due not so much to speculation or to giant-firm machinations as to the lack of research on technologies that would actu-ally increase food production (Kiers et al. 2008; Normile 2008a, 2008b). For years, research had gone into areas like genetic engineering that paid off hand-somely for the giant firms but did not actually increase food production much. The funding of the International Rice Research Institute, the world center for research on the world's most important staple food crop, peaked in 1993 at $44.4 million, declining to $27.9 million in 2006 (Normile 2008b). There are individual people who spend more per year on parties than that. By contrast, the largest firm in the world, ExxonMobil, had profits of almost $50 billion in 2008. That gives it plenty of political power, as well as the ability to maintain an enormous campaign of disinformation about global warming.

The big firms were not demons (as *Grain* implied) nor angels either (as the *Wall Street Journal* tried to insist), but were simply doing what they were supposed to do: make a profit. Nor was the market the villain (*Grain*) or the savior (*Wall Street Journal*). The problem was far wider. The immediate cause of the crisis was government-subsidized diversion of grain from food to biofuel production. Other causes included rapid increases in demand for both food and fuel. The latter raised the price of oil, essential to modern farming. Also, vast areas of prime farmland have either been desertified or turned to parking lots and suburbs in the past half century, and destroying the rainforests to make more cropland has not been an adequate substitute; the soils are less good, and the areas are remote. Finally, government policies in many countries—notably including the United States and also Mexico—have involved less and less help and aid to small farmers and more and more subsidies to the giant firms.

SUBSIDIES

Destructive activities would be countereconomic, in a wide range of cases, if they were not subsidized. Subsidies to oil corporations, including immunity from lawsuits over pollution, make it economic to use vast amounts of fossil fuel. Subsidies to fishermen make it economic to catch the last fish, when market forces would shut down essentially all fisheries long before the fish were gone; "commercial extinction" would come long before literal extinction. Subsidies to logging firms allow them to log impossibly steep slopes, bring low-value logs to market, and stop bothering with management for sustainable yield. Subsidies to developers and homeowners produce huge houses on sprawling lots. Local hunting that exhausts game is subsidized—in an "informal" but very real sense—when wardens and police fail to enforce the laws because of corruption or indifference. Almost every environmental problem I have ever looked at (and I have looked at a lot, in many countries) was subsidized, and few would have continued if those involved had had to turn a legitimate profit.

Worldwide, money spent on specifically conservationist activities is around $20 billion, but subsidies for uneconomic, throughput-maximizing production is at least $200 billion (Myers 2008)—a very conservative figure.

Some defend these subsidies on the grounds that they "create jobs." A more critical literature (see Chapter 6) holds that anything even approaching a free market would eliminate the profitability—and therefore the unfortunate presence—of unsustainable and destructive logging, farming, grazing, and so on. The most critical literature attacks these large interests head-on—an approach that leads to throwing the baby out with the bathwater. We need to stop unsustainable and destructive practices, not end all economic activity.

The most well-known example is fossil fuels. In the United States, even at current prices, they are notoriously underpriced, because of huge subsidies, tax breaks, and political advantages to oil corporations (Juhasz 2008). The ordinary folk thus use far more oil, gasoline, and coal than they would if it were priced at their true cost. Food production and transport, commuting to work, shopping at huge centralized stores instead of neighborhood markets, and indeed all modern, middle-class life depend on available gasoline. Subsidies to fossil fuel industries currently run around $15–$35 billion in the United States, depending on what counts as a "subsidy," and around $10 billion in Europe (Cleantech Forum online, posted Jan. 5, 2007).

In addition to direct subsidies, there are disaster relief funds (very generous for farmers and oil corporations), roads, agricultural research and development, extension services, rural electrification, low-interest construction bonds, sweetheart deals and specific federal favors to giant firms, the oil depletion allowance and other tax giveaways, and above all the near immunity from prosecution for pollution (see Johnston 2003, 2007; also following chapters). The oil industry in the United States has low-cost, long-term lease access to federal lands, and the Republicans keep up the pressure to open the national wildlife refuges and the offshore fields to oil drilling. Past experience proves that little or nothing will be done to make such drilling environmentally safe. Internationally, the oil industry is associated with every sort of problem from environmental devastation to failed states (Humphreys et al. 2007; Juhasz 2008).

Norman Myers noted in 1997 that market forces would set gas prices at $8 a gallon if environmental costs of using it were figured in. Indeed, it soon rose to $8 a gallon in Europe. Others have estimated the true cost of gas at several hundred dollars a gallon!

Fish are another case in point. Myers also pointed out that world fisheries absorb $124 billion in costs while producing only $70 billion worth of fish—subsidies again (see also McGinn 1998). Things have only gotten worse since (Sumaila and Pauly 2007; Worm et al 2006).

Inordinate consumption of gasoline and of fish is due more to artificially low pricing than to our sinful nature. Jeffrey Vincent and Theodore Panayotou, discussing Myers' position, pointed out that consumption could simply be made more economically sane by pricing resources properly, using them more efficiently, and shifting consumption to less "consuming" forms

(Vincent and Panayatou 1997); Myers later adopted this position. Myers notes that the microchip does take less material than the automobile—there is hope in the new technology (Myers 1997:55). He might have added that a small new hybrid automobile leaves, in its life span, about 3 percent as much of an ecological footprint as a huge gas guzzler of the type popular in 1960. On the other hand, one must note that the microchip has not replaced the automobile, and the computer in which the microchip is embedded is an enormous source of pollution.

Subsidization makes efficiency seem uneconomic. With present technology, only a minority of the energy generated by burning fuel is used to drive cars or otherwise do work; the rest is waste heat (though it is heartening to know that the percentage doing actual work has risen over the years). Only a small part of the logs cut in most lumbering operations are used; the rest are thrown away. In many logging operations almost the whole tree is used profitably, but these occur where markets and infrastructure are developed and subsidies do not totally distort the costs of production. Tropical forest logging observed by this author (and I have seen a lot of it, on every continent that has tropical forests) typically wastes well over 90 percent of the wood cut.

Backward technology and imperfect markets limit the value of trees. It is often almost impossible to get them out of the bush. If they do get to town, there are no proper sawing or curing facilities and no marketing facilities for quality lumber. In Mexico I have seen rare tropical hardwoods, worth thousands of dollars per tree in the United States, made into $10 railroad ties or simply burned where they lay (Anderson 2005). There was no way to get them to the American market.

Direct subsidies and expenditures for logging, public-land grazing, big agribusiness (as opposed to small farms), large-scale water projects, oil, mining, urbanizing wild lands, and pollution now runs about $39 billion (Myers 1998:xxi). The indirect subsidies—including tax breaks, possibly the most common and most pernicious form of subsidy—must be several times that but are impossible to calculate accurately. Mark Zepezauer and Arthur Naiman (1996) found that subsidies, tax breaks, and direct fund transfers from the U.S. federal government to the affluent total over $448 billion annually—and this without counting in secret matters, indirect subsidies, and many of the more obscure and hard-to-quantify tax breaks. This compares with $130 billion in welfare for the poor in 1996 (Zepezauer and Naiman 1996:157). Of course the sums would be much greater now, thanks to inflation, political victories by industry, and the Iraq war, which became (among other things) a subsidy to Halliburton, big oil, and the munitions firms. A comment on the level of investigative journalism in the United States is that I cannot find more up-to-date figures.

Agricultural subsidies in the United States reach 39 percent of farmers (this and what follows is from Grunwald 2007). Subsidies cost $189 per American. The top 10 percent of farmers receive $34,190 direct subsidies each; the bottom 80 percent get an average of only $704. Only the major staple commodities are subsidized by direct supports. Hence in North Dakota (raising wheat and other grains) 78 percent are subsidized and 70 percent are subsidized in Iowa, but in California only 9 percent are. Riceland Foods is the biggest single receiver: $541 million since 1995. Meanwhile, small farms (under $250,000 per year in revenue in current dollars) have declined 32 percent since 1982. Fewer than 1 percent of the work force in the United States is farmers. U.S. cotton subsidies depress world prices so much that ending them would increase Malian cotton farmers' incomes by 5.7 percent. Goldberg (2006) reports that subsidies cost developing nations $24 billion a year by this route.

No one in national politics wants to end this because farm states are generally swing states, and they tend to vote for whoever took best care of agriculture.

Most other major nations have something similar. The result is overemphasis on particular commodities, often those that use exceedingly heavy inputs of fertilizer and pesticides. Even Jonah Goldberg (2006), right-wing author noted for his robust support of large firms and their activities, noted that this is a boondoggle, partly because $1.3 billion goes to people who do not farm at all—rural landowners who have managed to tap into the system. And, though he is an opponent of "welfare," even he was a bit shocked to find that these subsidies are 50 percent more than all welfare payments in the United States.

Biofuels are another major boondoggle. They are competitive only because of subsidies (Loyn 2008). They have diverted so much maize, soybeans, and palm oil from the food chain that food prices have skyrocketed. Rising demand, chronically inadequate investment in agricultural research, and rising costs of petroleum also sent food prices up, but the biofuel push was critical in sending food prices through the roof in 2008. Subsidies had acted before to keep food prices low, so perhaps it balances out in the long run—but the chaos of the political subsidy economy is devastating to the world's poor.

Subsidies may be the biggest single problem for the world's environment. Worldwide, governments pay out at least $1.5 trillion a year for environmentally destructive activities (Myers 1998, 1999; see also Anderson 1996; Bovard 1991, 1996).

Big government is easily corrupted, coming to represent what it should be controlling. Farm subsidies, once sold as a way to save the "small farm," have mushroomed now that the small farm has been replaced by agribusiness. These subsidies are ecologically devastating, since they distort production processes by making heavy chemical use, wetland drainage, monocropping, and other destructive practices profitable when this would not be the case in a freer market (Anderson 1996; Bovard 1991; Myers 1999).

Western American water is notoriously a political issue, and the U.S. government (especially the Corps of Engineers and Bureau of Reclamation) have heavily subsidized the rich: farmers, developers, resort owners, loggers, and indeed anyone with money. ("Water flows uphill to money," we say in the West; see Baden and Snow 1997; Myers 1998; Wilkinson 1992.) Agricultural uses of water are very heavily subsidized in such dry regions as the Southwest and California. Even water for subdivisions is uneconomically cheap. Moreover, the government is quick to put in huge and unnecessarily expensive flood control systems for those same subdivisions as a subsidy to the developers. The developer is not forced to think about the matter.

When developers do have to think, they can come to some creative solutions. I have observed that many parks in my area of the world—the Pacific states—occupy floodways. There is always water to keep the parks green, and the parks provide places for the water to spread out and be absorbed. But such parks are found in areas where conservationists or planners objected to concrete-lined ditches (which actually make serious floods worse; personal research). Regulations that open up free markets for water, and provide recourse to flood-ruined suburbanites, are slow in coming, though the reform process has finally begun.

Those who get rich by polluting or overexploiting can use their power to obtain subsidies for their productive processes. Enterprises that are subsidized for the best of reasons lose the incentive to be efficient or nonpolluting. Just being exempted from pollution controls or cleanup rules is subsidy enough, but many heavy polluters get huge direct and indirect financial subsidies, removing still more of the economic incentives to produce efficiently (Anderson 1972; Bovard 1991; Myers 1998; Pye-Smith 2002). This abuse has become so flagrant that even Paul Wolfowitz, one of the most dedicated apostles of corporate power, has spoken out against it (Wolfowitz 2005). Direct subsidies include direct cash payments and tax breaks. Indirect subsidies include protective tariffs, infrastructure developments such as paved roads and port facilities, government-funded research dedicated to the industry in question, selective preservation and protection, selective law enforcement (very common in the Third World), and the like.

Development agents are apt to be particularly friendly to monolithic interests, both for financial reasons (usually rather less than moral ones; see e.g. Painter and Durham 1995, Stiglitz 2003) and for ideological ones (Scott

1998). The situation is bad enough in the First World, but the Third World is particularly hard hit. A vastly disproportionate amount of aid and development assistance goes to development of a tiny handful of commodities wanted by First World and East Asian countries.

In southeast Mexico, anyone who wants to start a cattle ranch (and thus destroy the local ecosystem) can get vast amounts of research and extension material—to say nothing of loans, subsidies, marketing help, and much else. But anyone trying to commercialize the superb and nutritious fruits and local crops of the region is out of luck (this, once again, from my Maya research; see Anderson 2005). The mamey, which produces hundreds of extremely nutritious fruit under appallingly harsh conditions, has never been studied, nor is it in commercial nursery-stock production (so far as I am aware). Chaya, a green leaf vegetable of the area, is so nutritious that a serving of it compares to a multivitamin pill, yet it has rarely been studied. (I was able to lure my student Jeff Ross-Ibarra into studying this plant, so at least a beginning has been done now; Ross-Ibarra and Molina-Cruz 2002. Much more is needed.)

Subsidies allow firms not only to expand and succeed, but make them more and more able and eager to provide "campaign contributions" and downright corrupt payments to the politicians who subsidize them (Pye-Smith 2002). They can thus win by "farming the subsidies" instead of by making better products (see, e.g., Eichenwald 2000; Myers 1998).

Polluting firms sometimes actually balance the costs of bribing politicians against the costs of cleaning up or innovating more efficient processes (Humphreys et al 2007). Certainly, paying politicians for favors, including safety from environmental enforcement, is often cheaper than correcting the problems and is a frequent recourse of firms and small producers in some countries (Ascher 1999).

Beyond subsidies are the primary-production firms that are outright state creations, like China's, Indonesia's, and Mexico's national oil firms. Other giant multinational firms are parastatals. Others are theoretically private but are so closely connected to the government that they are de facto a part of it, like Halliburton Corporation in the Bush-Cheney administration.

Another major type is infrastructure development that favors one sector, such as automobile roads versus rail and even zoning laws that force separation of workplaces and homes. Trade policies can favor one industry, one company, or one nation over another without being outright restraint of trade. Laws can favor one sector over others. Bans on "frivolous" lawsuits in the United States have included such things as outlawing all criticism of locally produced foods, even dangerously contaminated foods. Several states with major agricultural interests have passed such laws. While these laws are clearly unconstitutional (they violate the First Amendment) and do not survive court challenge, they have successfully scared the media, which now tends to avoid anything critical of contaminated food or pesticide residues on food. Thus, Robert Hatherill, a food researcher at the University of California, Santa Barbara, reports "my publisher stripped lengthy passages from my new book. Simply put, I was not allowed to disclose dangers inherent in some common foods . . . as well as over-the-counter medicines" (J. Robert Hatherill, "Take the Gag off Food Safety Issues," Los Angeles Times, p. B5, April 12, 1999). He was advised that the publishers would surely win any legal cases, but his book was not worth the money they would spend in court. Thus does flagrantly political and unconstitutional abuse of law silence the whistle blowers.

The saddest case occurs when subsidies corrupt industries started as environmentalist alternatives. The most extreme example is biofuel. Subsidizing maize (even more than before) and biofuel development has led to soaring food prices worldwide. This indirectly led to the destruction of the Brazilian forests for soybean planting, the Indonesian and Malaysian forests for palm oil, and other devastating development around the world. The end result has been to put rich SUV drivers into direct competition with starving Third World people (see, e.g., Paul Krugman's column in the New York Times, April 7, 2008). The environment lost; destroying forest produces tens to hundreds of times as much greenhouse gas as the biofuel saves. The losses in biodiversity are incalculable.

Government becomes one arm of the resource-transforming sector of the economy and loses its other functions. Extreme centralization in all things becomes the norm, stifling free enterprise as well as free speech and opinion. The economy, meanwhile, becomes based on resource extraction, not on production for actual use. Landlords, or resource lords, dominate the country; capitalists (in the strict sense of the term) do not. (For the details of how this works on the ground, see Ascher 1999; Bunker and Ciccantell 2005; Harich 2007.)

This sort of thing is often called "neoliberalism" (Harvey 2006). However, this term is based on a series of errors. Neoliberalism, a term that has been used since the 19th century, applies correctly to a philosophy of free trade, including free international trade. This correctly identifies the pressures on Third World countries to abolish protectionism and to run down their public infrastructures, notably health care and education. However, it miscalls everything else. The fundamental dynamic has been for enormous and rapidly increasing subsidy of giant First World firms and the larger communist statal and parastatal firms. This includes extreme pressure, by the nations involved and by the World Bank, IMF, and WTO, to open Third World countries to these firms. Obviously, this is not free trade; free trade would mean that subsidies, trade biases, and government policies benefiting one particular firm or industry at the expense of others would cease to exist.

The WTO has, in fact, very selectively enforced this. They are perfectly willing to take on tiny industries, but taking on Big Oil or Big Pharma is not even mentioned.

There are two iron laws of producer subsidies: First, they always end by benefiting the rich, usually at the expense of the poor and always at the expense of the general welfare. If the subsidized do not start out rich, they soon become rich, because of their advantage.

Second, they usually benefit the backward sectors of the economy at the expense of the forward-looking ones. The forward-looking ones are doing fine on their own and generally prefer to spend their money on actual entrepreneurship than on lobbying and bribing for special favors. Moreover, if a favored industry does not start out backward, it soon becomes so because the subsidy removes the incentive to progress. It finds it can do better by jacking politicians for subsidies than by innovating. This process has been well described (e.g. Bovard 1991; Myers 1998).

We have no idea of the economics of this behavior worldwide. In the United States, we can find out something of the level of contributions to politicians, but even in the United States the real financing of campaigns is less transparent. Campaign reform laws have dried up the naked influence of big oil, big autos, and big agribusiness—the top donors in the 1990s—but have probably driven influence underground rather than eliminating it. Other nations do not have even America's weak transparency and disclosure laws; comparable information is disturbingly unavailable. One wonders what the record would look like in Mexico, Colombia, or Gabon.

Thus Pye-Smith (2002), looking at the world situation, calls for an end of subsidies. This is perhaps too much; we may need to support food production, energy generation, and other necessary production. The problem comes when it leads to distorted priorities, especially overuse of polluting products. It may be, however, that large-scale subsidization is so inherently flawed that it cannot do anything but harm.

ACCOUNTING

The situation is not improved by the tendency of governments and their economists to value only direct production—not the environment's services or the costs to the environment of productive processes (Daily 1997; Repetto 1992). It is clear that one problem with the world environmental cause is the lack of sane cost-benefit accounting in these matters (Anderson 1996). The global warming fiasco, when governments, especially the Bush administration in the United States, listened to oil and energy company public relations men rather than to scientists, has been an extreme case.

In other cases, there is a more pernicious and harder-to-address failure to account for "nature's services" (Daily 1997). The air, water, and soil of the world absorbed our pollution for millennia but cannot do so any longer. We have treated fresh water and good soil as free goods, or at least as lavishly available ones.

We are poorly informed about social costs. How much damage to local communities is being caused by huge logging projects, pollution of public parks, or freeways punched through residential neighborhoods? Everyone knows these things are socially damaging, but no one can cost them out adequately.

The lives and livelihoods of the poor do not count for much. A farmer in Madagascar earning $300 a year (a relatively high income there) is making about 1 percent of the average income of an American. In turn, the average American earns approximately 0.000001 percent (sic) as much as Bill Gates. Bill Gates' income is greater than the gross national product of Madagascar.

Yet, the subsistence interests of the poor do not exist, so far as economic accounting is concerned. The World Bank, WTO, and giant multinational NGOs cannot see subsistence production. They do not take it into account when assessing the damage done to local communities by big dams, urbanization, population growth, or replacing inhabited forests by agribusiness. Yet, if my Mexican research is typical, subsistence farmers produce goods worth about $1,000 per year per person. I suspect this figure is indeed typical, in which case the world's billion or so rural poor have lost, or are losing, a trillion uncounted dollars in goods. (I am aware of lesser estimates, but they are based on looking at already damaged ecosystems, not the ones in Mexico, the Northwest, and Malaysia that I saw in good shape.)

Comparable misaccounting is universal (Ascher 1999; Dichter 2003; Stiglitz 2003). For instance, in the case of large dams noted above, the damages to the people displaced by the dams are either undercounted or not counted at all (Scudder 2005). Most or all large dams, and certainly many projects involving deforestation and other land conversion, would not be economically viable if this accounting were done.

I extended this model in the late 1960s (Anderson 1996, 1999 [orig. 1968]) to take account of political power and cultural rules. Individuals will naturally use any political leverage they have. The beneficiaries of externalization will advocate laws that institutionalize such unfair cost-benefit systems. Exploitive interests take advantage of this to sell short-term strategies (cf. Taylor 1989).

Economics thus exists within a surrounding shell of politics. Polluting firms or fishery-depleting fish corporations are often protected and subsidized by governments.

However, I do not mean to blame corporations and governments alone. Politics, in turn, exists within a surrounding shell of culture. Ethics, cultural wants, religion, and other cultural institutions all create the context in which politics happens. Fish-loving Japan, for instance, is more guilty of overfishing than the United States, whose populace does not care enough about fish to insist on access at all costs. Conversely, however, and for the same reason, the Japanese are more concerned about fish-killing pollution than is the United States. In such cases, culture is more important than economics or political power.

Moreover, bureaucratic structures take on a life of their own—they become "emergents" in social science jargon. Accounting failures occur especially under two conditions: when higher-ups isolated from the scene take over the decision-making functions and when the decision makers are not accountable or can somehow dodge the accounting.

Of course, cost-benefit accounting and all other economic questions concerning natural resources are difficult to do with anything like perfect accuracy. Fortunately, perfect specification of all benefits and costs is not necessary. Even an approximation to adequate cost-benefit accounting would stop much environmental degradation. Big dams and other government pork-barrel projects often do not look good even in the cost-benefit accountings of their own parent agencies (Scudder 2005).

Experts advise for a very careful approach, looking to the worst-possible-case scenario, but governments all too often go with the rosiest predictions (Ludwig et al. 1993; Rosenberg et al. 1993). At first this may be a reasonable mistake. However, as Ludwig, Hilborn, and Waters point out, when bad years hit, "the industry appeals to the government for help; often substantial investments and many jobs are at stake. The governmental response typically is direct or indirect subsidies . . . their effect is to encourage overharvesting . . . The long-term outcome is a heavily subsidized industry that overharvests the resource" (Ludwig et al. 1993:17). They go on to document overestimated resource quantities, compromised regulations, and all the other supporting cast of political biases. They counsel extreme caution in resource use planning—but provide no advice on how to institutionalize it, except to warn governments to listen to cautious majorities, not rashly optimistic minority positions, when experts disagree. One can always find "experts," usually working for the companies in question, who will say that any level of overuse is fine.

ECONOMIC CURES

There is now a great deal of literature on better environmental accounting (e.g., Baskin 1997; Daily 1997; Daly and Cobb 1994; Repetto 1990; Roush 1997). Nor is it only liberals that argue thus. For a libertarian overview that balances Daly's broadly "liberal" view but is equally valid as a conservation strategy, see John Baden's work (e.g., Baden and Noonan 1998). The values of medicinal plants from rainforests alone probably run far into the millions (Balick and Mendelsohn 1991). Robert Costanza has estimated the value of natural systems at $33 trillion (Roush 1997). Some find this high; I find it far, far too low, since Costanza has—among other things—undervalued the potential that new crops would have if people were really reduced to using them by (let us say) a worldwide wheat blight. He has argued for "a new tax on depletion of natural capital such as wetlands" (Roush 1997:1029). Arrow et al. (1995) have indicated that the particular mix of inputs and outputs used in an economy are important (how much pollution is generated, for instance) and that policy should be set accordingly.

Economists have developed ways to deal with this. Mancur Olson (1965) thought of "side benefits": parties, rewards, and other prizes to the cooperators. This is demonstrably not effective enough. Game wardens, like other security guards, have to be willing to risk their lives to protect their charges; no side benefit can make up for that, unless self-respect is a "side benefit."

Institutions can reduce transaction costs (North 1990) and thus make such things easier. Institutional economics has a good predictive record (North 1990; and Schultz 1968, an early text of this school, is still very valuable). However, this does not really solve Olson's problem. People motivated by nothing but self-interest would still cheat. All one has to do to be convinced is to look at any corrupted government agency. The world has many institutions, and a substantial percentage of them show just the sort of rip-off mentality that Olson modeled. Yet, obviously, some institutions do work. Game wardens in many countries have died for their parks, just as soldiers and even gangland toughs have so often died for their comrades.

Economists postulate "infinite substitutability," a theoretical assumption that any resource could be replaced by another resource. However, air, water, and soil turn out not to be substitutable. People have to have them. So far, we are also unable to create species, let alone air and water, and we depend on nature's services to provide them.

Economists have, accordingly, striven to find ways of creating sustainable and efficient economies (Daily 1997; Daly and Cobb 1994; Murphy 1967; Van Dieren 1995). It has been widely argued that we must change our basic outlook on resources and economics if we are to accomplish this. Some are hopeful that small changes, in the direction of freer markets and more accountability, would do the job. Others argue for far more radical solutions, including the downfall of capitalism. This leaves a vast middle ground, in which (I think) most people find themselves. I believe the best middle ground is one on which free (i.e., relatively free) markets can function, but within limits set by a morality of responsibility, backed up by legal institutions (cf. Friedman 2005).[1]

NEEDS FOR SOCIETY

Fortunately, in spite of political extremism on the left and right, conservatives as well as liberals have provided many alternatives. Economic cures for environmental problems have been proposed and locally tried with great success. Most important has been making polluters pay the costs of cleanup, making developers pay the cost of damage to developed lands, and otherwise specifying the costs of production on those who benefit directly. This shifts the cost-benefit calculus, making it more realistic (Murphy 1967). Another system that works with great success is comanagement: governments, firms, or conservation organizations working with people on the ground who are engaged in production (Pinkerton 1989). The local people regulate their behavior; the outside agencies provide advice, mediate disputes, coordinate efforts, and sometimes set standards. Improved government responsiveness has worked well in many cases. Creating a genuine free market, instead of the system of subsidies to production that exists in most countries, could theoretically benefit the environment (Baden and Noonan 1998; Anderson and Leal 1991, 1997), but it cannot always protect people from long-term consequences of immediately profitable actions. There are some things governments have to do. Protecting people is one. The government has to protect, and that involves stopping people from genuinely damaging activities like pouring poisons into the air and water just as surely as it involves stopping people from using guns and knives to murder.

Justice is another. Clearly, *the rich countries and regions must bear some of the costs* if anything like justice is to be served (Lizarralde 1998), and as we have seen, justice is essential in these matters. This is especially true if the rich were colonial powers and the poor were their subjects. On the other hand, more than one tropical nation has used this discourse to excuse its own failure to make even a minimal attempt to control environmental destruction.

This is not to say that governments should be held to a sort of ecological blackmail, forced to pay to save endangered resources (especially on public lands leased to private operators; Williams 1993). Governments naturally compensate, or *should* compensate, for taking private land (but *not* for redesignating public land) for national parks; there has to be a similar worldwide system for saving habitats and biodiversity. The principle is, those who share benefits should pay the costs of maintaining them; on the other hand, those who receive damages should *not* have to pay blackmail or protection money to avoid the same. If upstream users are polluting the river, the downstream users should not have to pay the upstream users to make them clean up. Civil damages are to be stopped, not subsidized.

One reform with interesting possibilities is the institution of severance taxes on resources. Such taxes have been proposed occasionally by economists, but the idea is so "radical" in the context of 20th-century economics that few people have followed it up. However, Mason Gaffney (Gaffney and Harrison 1994) and David Rodman (1999) have made a case for this. (I am grateful to my colleague Mason Gaffney for pointing out that he, and some others, have been seriously advocating forms of this for years.)

We already have the idea in place: "stumpage fees" for logging national forests, barrel taxes on oil, and grazing fees on public land leases are all there. The problem is that (at least in my experience, largely with forestry and grazing; though see Hedrick 2007) they do not even begin to cover costs. They partially offset the subsidies, but only partially. I would argue that they should be raised to not only cover the direct costs to the government of making the resource available, but also to cover the overall costs to society of damaging the environment. They should be dedicated entirely to fixing the environments from which they are collected, not liberated to general funds.

In an ideal severance-tax world, there should be large taxes on raw materials production and increasingly smaller ones for each resource transformation or reuse. Instead of paying people to destroy the ecosystem, we should be charging them upfront for the total costs (in so far as known—there is simply no way to figure in the unknowns). Surely a tax on ecological severance is the fairest tax. An income tax penalizes work; a severance tax penalizes waste. Resource taxation should not only cover the social costs of production—it would produce all needed government revenues, from defense and crime control to delivering the mail. Property taxes should be levied on only those people who "do something" with their property. We should never, ever tax property that is left undeveloped; we should tax property according to how much is done to it. As it is, property taxes (at least in places I live) are often assessed according to "best use," so that an undeveloped piece of property is taxed as if it were a housing development. We should not tax it at all; we should impose taxes as it is developed, according to the value of the development done.

The "fair tax" advocated by conservatives—a national sales tax or value-added tax—has some of the right features but penalizes the ultimate consumer, often a less-than-affluent individual doing his best. Instead, *the actual original producers should pay the taxes.*

Jose Drummond (personal communication) has pointed out to me that an environmental severance tax could be hard on small subsistence producers. I would advocate a progressive tax, but, in fairness, even the smallest of grazers, fishers, and others who are really damaging the land should have to pay. Traditional subsistence cultivators such as the Yucatec Maya do so little net harm to the environment that they would not pay in a reasonable tax system.

The obvious problem with such a scheme would be the need for near "perfect information" to determine levels of abuse. However, we have enough information to make it work well enough in many cases. We could make a start and work out from what we know.

Forest Reinhardt's excellent guide for environmentally conscious business managers, *Down to Earth* (2000), makes very clear the exquisite sensitivity to different legal and governmental systems that characterizes the private sector. Even the threat of regulations, and even the very mildest regulations, are enough to change profoundly the behavior of firms. Most important of all, even the slightest change in consumer spending patterns can produce enormous reactions in the business community. The thoughtful reader, especially the careful reader of Reinhardt's cases, is struck by how much power the individual has and how very little effort it would take to reverse much of current behavior. On the one hand, even a very small effort would fix the world. On the other, an equally small effort would eliminate what little protection we have.

If decision making is removed from the local level, accountability is lost. The farther it is removed, the more it becomes the task of faceless bureaucrats who have nothing to lose from wrecking a resource.

Yet the higher levels are now, inevitably, about equally important. Governments regulate economies and provide infrastructures: roads, weight and measurement standards, courts, flood control, defense systems, postal services, school systems, and electric grids. Governments structure economies by defining property rights, patents, licensing, trading rules, full faith and credit clauses, and thousands of pages more of civil legislation. Any of these can be socially constructed in an infinite number of ways.

The opposition of capitalism and socialism crosscuts an opposition of local grassroots control and top-down, centralized control. The goal is to have central administrations do only what must be done at a higher or general level. This is simply good management; it has been part of management theory since Shen Pu-Hai and his "rectification of names" (i.e., specifying the right level for making decisions) in ancient China (Creel 1972).

We are, thus, left with the question-begging conclusion that mixed systems should be best. Scandinavia and the Low Countries, with their almost perfect equality in economic clout between private and public sectors, are currently the most environmentally responsible countries. In fact, mixed systems are almost inevitable. Only they can bring together local and wider concerns in one system. In any case, neither the "free market" nor the socialist utopia are really imaginable—to me, at least—in the contemporary world.

Another arena where a balance is difficult to find, but necessary, is the relationship of individual to system. Individualism is valuable, but it has to be tempered by a strong sense of social responsibility. (Again, northwest Europe does very well at this.) The "selfish" user is not a social "individual," but a parasite. Individuals define themselves through interaction, and interaction must be the node for examining environmental action.

All levels of the hierarchy must be able to prohibit destructive behavior and to protect life and property, but no level except the most local one possible should be able to take economic initiative.

I would advocate that power return to the people, by which I mean, explicitly, the vast majority of individuals. A small party or single individual that claims to "represent the people" is the opposite of what we need.

"The vast majority of individuals" does not mean the tyranny of the majority. Local people deal with local situations, while the general majority protects wider interests, as when migratory birds are managed across national boundaries. Worries about the future of local autonomy, however, lie in the progressive globalization of our problems. Global warming due to greenhouse gases, depletion of forests due to international timber marketing, and competition for fish by fleets from many nations all present us with problems more intractable than saving migratory birds.

An ironic part of environmental politics has been the cost of success. In the 1960s and 1970s, scary, but perfectly true, accounts of world hunger and poverty frightened politicians into correcting the worst problems of pollution, population growth, and especially world hunger. Sober predictions of mass famine (e.g., Ehrlich and Ehrlich 1968) would have been perfectly correct, if they had not terrified politicians and foundations into making all-out efforts. These included the Green Revolution (see Lowell Hardin's personal account of the Bellagio conference in 1969; Hardin 2008). Nothing makes me happier than seeing a dire prediction effectively defanging itself, but it does have the disadvantage of giving the anti-environmentalists a chance to gloat about "Chicken Little" and "crying wolf." (Anti-environmentalists seem to read at the six-year-old level.)

Needs for the future include full and honest cost-benefit accounting; elimination of subsidies for inefficiency; and markets, very likely freer ones, but within a moral shell. There are many ways to achieve such goals, but all demand solidarity, social responsibility, and mutual trust. The reasons are obvious: Without agreement—a Hobbesian social contract—people will not be able to prevent competition that shortens the short-term to the vanishing point.

What, then, would a green economy look like? In my personal vision, it would be a much freer market than anything today. That would not mean that giant corporations are allowed to do anything they please at taxpayers' expense, while individuals face more and more disadvantages (the "neoliberal" pattern). It would mean that the playing field would be leveled in terms of opportunity. Subsidies for environmental destruction would be ended. Concentration of power would be discouraged by progressive tax rates and antimonopolizing rules. Economies of scale are often genuine—no harm in that. The trouble is that they are often achieved only because the giant firms have an easier time passing on the real costs of doing business as "externalities." We need to let more perfect markets work, to reward innovation, efficiency, and good accounting, and to squeeze out inefficient dinosaur industries.

It would be a highly diversified economy. The emphasis would be on efficiency and on narrow niches, rather than on a few giant firms selling a few uniform products worldwide.

It would be an economy of services, not material products. Material products would be minimalist: They would use the least material possible for a given purpose and would, ideally, serve more than one purpose.

It would be an economy in which the government served mainly to protect and to limit. The government would be funded largely or entirely by severance taxes, not income taxes.

It would be an economy of local control of resource use—overseen by wider polities, with (to repeat) a protective function. They would act to limit local abuse, not to force locals to destroy their environment for distant profiteers' benefit.

It would be an economy in which consumers cared about quality at least as much as quantity and one in which they did not have to pay inordinately for the choice. (The current world of giant subsidized corporations makes mass-produced junk artificially cheap, for reasons stated above, while quality production is at multiple disadvantages.)

A utopian world based on these principles may not be totally impossible. More to the point is the fact that all these are process goals. Any progress toward them, if at all well managed, would be an improvement. One can think of nightmare scenarios in which attempts at transition to such an economy would produce a monstrous mess, but such scenarios are based on people doing it wrong. We are at least spared the nightmare of subsidized production: It is worst when done best.

We might, then, do the following:

- Minimize subsidies to production. Subsidies have a place: They should be for innovating new, more efficient ways of using and saving resources, or they should be for *non*consuming.
- Raise the biggest share of public money through taxes on consumption. A severance tax on resource use would be imposed. Taxes on transformations of power and materials would be imposed. The tax rate might well be adjusted to hit especially hard the purely wasteful uses of material, like overwrapping (extra packaging). Heavy fines on polluting and waste would raise a good deal of the rest of the national wealth.
- Set tax rates to maximize efficiency.
- But, also, I would keep the progressive income tax. It is needed to help level down the overly powerful. In any case, the marginal value of a dollar (or even of a percentage of total wealth) to a rich person is less than the marginal value of a dollar to a poor person.
- Give tax breaks for conservation and preservation of natural or more-or-less-natural environments. This would need to be done with caution, however, for obvious reasons.
- Invest very heavily in education, science, and research of all sorts but (of course) notably in field biology and related areas.
- In the end, people's revealed preferences—what they spend their money on—must be ecologically sound. If I drive to the mountains to look at the scenery, I may personally prefer the mountains, but my revealed preferences are for a car and gasoline. Ecologists who constantly jet from conference to conference are revealing a preference for wasting airplane fuel, not for the cause.

5

Environmental Justice

ORIENTATION

All that has gone before leads directly to the wider issue of environmental justice. The poor have disproportionately borne the burden of environmental deterioration. The number of people now in desperate poverty and want is greater than the total population of the world in 1900 (see Collier 2007). Bland statements that environmental concerns are overblown, or that the gloomy prophets of the 1960s have proved wrong, are not easy to maintain when one travels in rural Madagascar or Bangladesh or China or a hundred other nations. Even the United States, in spite of its riches, has vast areas where poverty and want occur. Many Third World countries have few people *not* in such conditions.

My contention, based on research in many countries around the world, is that social and environmental injustice is not only basic and central to the world environmental problem but is the most important single cause of it. The rich pass on the real costs of production and consumption as "externalities." These impact and ruin the poor and many of the not so poor, who cannot do anything about the situation. If this were stopped, either by international law or by the less affluent getting some political recourse, world environmental abuse would be largely stopped. Therefore, we need to examine in detail just how the ordinary, nonaffluent people of the world are losing out.

The billion people at the bottom of the world's socioeconomic scale (Collier 2007) suffer from multiple problems. They have limited access to water—often *no* access to safe, clean water. They have almost no direct access to natural resources, even where their grandparents were subsistence farmers making a good living from the land. When they can access resources, they are hedged with ever-increasing restrictions on use and on their traditional management systems. They have little health care. They are faced with soaring costs of water, fuel, and materials as these run out. They are faced with soaring costs and falling availability of nutritious food, while low-nutrient, highly processed food becomes more and more easily available. They are rapidly losing the accumulated environmental knowledge of millennia, as displacement, poor education, and (frequently) endemic violence increasingly separate them from opportunities to teach and learn.

They are paying—simultaneously—the real costs of deforestation, over-fishing, farmland loss (to erosion or urbanization), water waste and pollution, air pollution, and every other sort of environmental degradation. These externalities interact; pollution ruins farmland and fishing (and makes the fish dangerous to eat), deforestation leads to soil erosion and removes forests that blot up some pollutants, and environmental degradation leads to more competition of all sorts for what is left.

Most of them live in more or less authoritarian regimes, and even those living in democracies are too poor and marginal to have much serious hope of recourse. Typically, they belong to groups suffering from discrimination and prejudice. They are not only last hired, first fired but also the last to have their environmental views considered and first to be affected by pollution and environmental deterioration. They can do less and less to affect their increasingly difficult fates.

They are not considered by the elites. Often, as Kristin Shrader-Frechette (2002:34–36) points out, cost-benefit accounting simply lumps poor and/or heavily impacted losers in with the affluent winners and assesses benefits to the whole nation as if everyone benefited equally—even when the heavily impacted share few or no benefits but a disproportionate percentage of the costs. This is equivalent to counting the costs and benefits of a burglary, averaging them out, and declaring the whole thing a wash and therefore no problem.

A similar problem concerns the livelihoods of the poor even when these are taken into account. A cost of $1,000 each to a million poor people is "only" a billion dollars—a tiny fraction of the annual income of Bill Gates or Warren Buffett. It makes every bit of economic sense to displace the world's poor for the benefit of a very few rich. As the rich get richer and the poor get relatively poorer, this spirals higher and higher. Whole nations can be sacrificed for the benefit of one or two giant corporations, and the cost-benefit accounting appears perfectly reasonable. If there is a tiny net gain, "economic growth" and "development" are served; total GDP is increased. The cold figures do not deal with the fact that millions are worse off and only a tiny group is better off. This sort of disparity is rare in industry and in urban economies, where benefits are spread more widely, but is absolutely typical in the world of primary production by giant multinational firms (Bunker and Cicantell 2005).

This is the real face of environmental injustice. It disproves the claim that environmentalism is an elite or antihuman concern. Quite the reverse: Those who dismiss environmental concerns dismiss those one to two billion suffering people.

America recently received a dramatic lesson from Hurricane Katrina. It destroyed most of New Orleans and displaced 300,000 people, most (though far from all) nonaffluent persons of color (see, e.g., Healey 2006). This was not unexpected. It followed decades of consigning the poor and nonwhite to low-lying ground. Developers had steadily compromised the safety of these vulnerable populations by construction of ill-planned canals and other facilities that served major corporate interests. The protective marshes that used to block hurricane surges were largely destroyed by channel cutting, oil and gas drilling, and flushing silt out to sea instead of allowing it to recharge the marshlands. Previous hurricanes (equally strong or stronger) had failed to destroy the city only because the coastal marshes were still in place. A hurricane-driven catastrophe was therefore fully expected. Detailed forecasts of it and plans for it were made over the years, climaxing in a full-scale simulation in 2004 (Fischetti 2006; Reichhardt et al 2005; Travis 2005).

However, nothing was done to implement the plans. Nor was anyone surprised when the poor and nonwhite—and even the not-so-poor working class—of Louisiana paid the costs. For decades, Louisiana has held the dubious distinction of being the poster child for environmental injustice. Dozens of polluting firms and land-consuming development interests have gravitated to the state precisely because of its tolerance of passing on the costs of production—pollution, flooding, harm to property values, and the like—to the less powerful communities (Roberts and Toffolon-Weiss 2001). What appeared to be an "act of God" turns out to have been an act of irresponsible humans.

The same can be said of Hurricane Mitch's impact on Honduras (Barbara Anderson, unpublished research, including extensive interviews of displaced peasants in Honduras; see also Stonich 1993). Peasants displaced by cattle ranchers had migrated to the towns and cities, where they were consigned to steep slopes to live. These slopes, cleared of protective forest and covered by squatters' houses, failed, leading to the deaths of many of these migrants and the dispossession of thousands more. Like Katrina, Mitch was a human-made catastrophe.

Most relevant research and advocacy has been in the area of pollution. This is especially true of action explicitly called "environmental justice." Similar activity in dealing with overfishing, overlogging, soil erosion, and related concerns, such as "sustainability," is also common but seems often termed "environmentalism" rather than "environmental justice." However, the link between sustainability and social justice is obvious and gets due attention (Brown 2003).

This has much to do with the rise of right-wing governments in many areas. However, it is also a natural evolution within environmentalism. Until the 1990s, environmentalists could say that the problem was simply one of people using resources and the solution was to stop them. By the 1990s, two problems emerged with this simple paradigm. First, people need to consume, and cutting resource use across the board would devastate the poor. Second, it is now clear that the problem is not so much overall use but inefficient, wasteful use. Both these problems are matters of environmental justice.

RISKS AND BACKGROUNDS

Nonwhites in the American South, for instance, bear a disproportionate and increasing share of the burden of dealing with industrial pollution and land degradation (Bullard 2000; Roberts and Toffolon-Weiss 2001). Overfishing hurts primarily the people who used to depend on fish as a cheap, readily available source of protein.

Even without deliberate intent to oppress, costs automatically get passed on down the socioeconomic system to the poorest and least powerful. This was semiseriously advocated by Lawrence Summers in an infamous memo (quoted in Liu 2001:i) proposing that toxic wastes be exported to poor countries because the lives and livelihoods of people there were worth very little. Even if we were prepared to abjure justice to that degree, Summers' proposal, if enacted, would remove all incentives to clean up. Waste disposal would remain so low in cost that there would be enormous counterincentives not only to cleanup but even to efficient production. The moral or efficient firms (if any) would not be competitive with the immoral ones. This is, in fact, happening today because of the export not of pollutants but of polluting industries. With heavy polluters relocated to countries like China and Mexico, where enforcement is minimal, the world economy is being subjected to this huge disincentive—a market failure of unprecedented proportions.

One corollary of this is that economic justice is economic wisdom—often the best course for the economy.

The vast majority of environmental justice sources deal with the problem of pollution, and much of the literature concerns the southern United States. This is in part because of the leadership of Robert Bullard, whose charismatic and path-breaking work (Bullard 1994, 2000, 2005; Bullard, Johnson and Torres 2000; see also Agyeman et al. 2003 and Shrader-Frechette 2002) has been concentrated in that area. Others focused on the political side have produced important studies (e.g., Bryant 1995; Checker 2005; Pellow 2004; Pellow et al. 2002; Sachs 1995; Szasz 1994). The title of one, *Deceit and Denial: The Deadly Politics of Industrial Pollution* (Markowitz and Rosner 2002), sums up a great deal.

Bullard, Checker, Shrader-Frechette, and others have linked the issue with civil rights. America's greatest concentration of polluting industries developed in black rural communities in Louisiana, producing a "Cancer Alley." Native American reservations have been exploited for uranium mining and toxic waste dump locations.

Moreover, the vast majority of research has been done in industrialized countries, especially the United States. However, the worst pollution problems are in developing countries. In China, huge chemical spills and dumping have recently made whole large rivers unusable (Economy 2005). India suffered the famous Bhopal incident, in which a Union Carbide plant broke down, killing thousands. Such catastrophes stand out against a background of constant "small" incidents and releases, which may ultimately kill and injure many more. Mexico has its border maquiladora, and its huge concentration of lead smelting in the Torreón area, which has delivered very large doses of lead dust to the children in this urban area of 2,000,000 people (personal observation).

One important cause of pollution-related death is pesticide overuse on Third World farms. However, statistics on this are notoriously unavailable or underestimated, so we do not know how extensive the problem is (see, e.g., Wright 2005); for one thing, ingesting insecticide has become a standard method for suicide in many countries, and it is hard to break out suicides from accidental deaths.

Linking environmental justice with minority rights is also a problem throughout the world. Anthropologist Barbara Rose Johnston has devoted a lifetime to pressing for consideration of the human rights of local indigenous peoples suffering from environmental injustice (Johnston 1994, 1997). Others have more recently joined this effort.

In view of the pollution issues, the Clinton administration innovated an environmental justice office within the Environmental Protection Agency. President Clinton's Executive Order 12898 of 1994 ordered federal agencies to consider and deal with environmental justice issues—not limited to pollution. Agencies were directed to identify and address "disproportionately high and adverse human health or environmental effects of its programs, policies, and activities on minority populations and low-income populations" (U. S. Government, EPA, 1994). In 1998 the federal government produced final guidelines for EPA analysis (Totten and Dickerson 1998). The

Millennium Ecosystem Assessment (2005) also stressed the need for environmental justice. However, the Bush administration turned away from these agendas; not only was there no significant legislation, but enforcement of the earlier laws was drastically reduced.

Environmental justice has also been of concern in forestry (Agrawal 2005, 2006; Guha 1990; Salazar 1996), mining (Gedicks 2001; Martinez-Alier 2001), biodiversity (Brechin et al. 2003), grazing (McCabe 1990, 2003; 2004; McCabe et al. 1992; Sheridan 1998), food security (Gottlieb and Fisher 1996), and plant use (Zerner 2000). Even recreation sites reveal the pattern (Cordell and Tarrant 1999). Rich and politically "connected" areas have far more and better access to recreation sites than poor and minority areas.

Meena Palaniappan et al. (2006) focus on access to clean water. Water is second only to oxygen among imperative needs of the human animal. There is no substitute for these needs; their case is the antithesis of the "infinite substitutability" basic to neoclassical economic theory. All traditional moral and religious codes concern themselves with water. It seems hard to believe that anyone could actually deny water to those dying of thirst, but current privatization of water and consequent skyrocketing prices do just that. So do governmental plans that waste water on a huge scale for the benefit of the rich. Big dams are the main example (Scudder 2005), but there are countless others, from development of exclusive suburbs to irrigation projects that benefit only a few large farmers. Few, if any, areas of environmental justice are as obviously a matter of concern as this one.

RESOURCE JUSTICE

A particularly difficult situation today involves large-scale water management (d'Estree and Colby 2004; Gleick 2004; Kobori and Glantz 1998). Much of the world is short of fresh water, and this situation is becoming rapidly worse. Most of the world's major dry-country rivers no longer reach the sea except in flood times. Examples include the Yellow River of China, the Nile, the Colorado, and the Rio Grande (Rio Bravo). Everywhere, supplies of safe fresh water are rapidly decreasing because of pollution, evaporation increase (due to big dams, irrigation, global warming, etc.), and exhaustion of "fossil" water in aquifers. Demand is rapidly increasing because of population growth, economic growth, and local factors. An example of the latter is displacement of agriculture by urbanization. Agriculture relocates from the best farmland, increasingly becoming urbanized, to marginal lands that often need heavy irrigation. D'Estrée and Colby (2004) examined a variety of plans and outcomes and evaluated the justice of the outcomes.

Notable are the cases in which an upstream nation is taking much or all the water from a river or could do so. An extreme case is the Colorado, almost all of whose water is taken by the United States. The river is completely dry by the time it becomes an entirely Mexican river. Turkey now takes so much of the Tigris and Euphrates that Syria and Iraq are seriously impacted. Ethiopia and Sudan could dry up the Nile, leaving Egypt's 60,000,000 people to perish. Cases such as these have led to serious fears that World War III will be about water (Gleick 2004). These cases display the extreme urgency of the need for environmental justice.[1]

Even global warming is harming the poor differentially. Global warming is caused, in large part, by release of greenhouse gases due to burning fossil fuels and to clearing of land (leading to breakdown of vegetable material). The leading releasers of greenhouse gases are the United States and China. These nations benefit. However, global warming has led to drying of Africa (Gedney et al. 2006; ironically, the United States and China are both getting increased rainfall, though not in their dry areas). Africa's droughts have led to thousands of deaths from starvation, shortage of potable water, and related causes including disease (see, e.g., *Sierra* 2006). *Sierra* magazine concludes that "Global warming is an environmentally unjust calamity" (*Sierra* 2006:13).

Often, local people were making highly efficient use of all resources, but they are displaced by outside interests that practice "rape, ruin, and run" strategies. Tropical forest logging has been notorious in this regard; local groups that had been using the forest sustainably for millennia are displaced by fly-by-night logging corporations that destroy the land. (I have seen this at first hand in a dozen countries; see Bunker and Ciccantell 2005.)

In southern Mexico, local indigenous people intensively and sustainably use almost every conceivable resource in a tract of forest but were sometimes displaced by government plans that cut down the forest for far less productive systems, or even just to get rid of it—to "open it up for agriculture" (Anderson 2005). Forests and forest diversity have been maintained—and may even have been created—by Maya management (Anderson 2005). Yet Maya have often been displaced by park and reserve projects (Anderson, unpublished research; Faust 1998 and unpublished; Haenn 2005). The Maya then lose their incentive to manage the land and become poachers, sell the land off, move away, or simply neglect what they once kept in superb shape (Anderson 2005; Haenn 2005).

Similar problems are common in southeast Asia, where forests essential to the survival of local communities have been cut down to produce disposable chopsticks, disposable construction siding, toilet paper, and other trivial uses. Indonesia has been almost totally deforested for these purposes or for conversion to agriculture—mostly plantation agriculture that produces products for export and benefits only the rich. Those displaced merge into the great pool of dispossessed persons and refugees that crowd the globe in the 21st century (Bender and Winer 2001).

DISPOSSESSION OF INDIGENOUS PEOPLE

Slash-and-burn cultivation has notoriously been subject to attack by governments because it "destroys forests." Sometimes it does (Terborgh 1999). However, sometimes it creates or maintains forests (Anderson 2005; Fairhead and Leach 1996) or has at least coexisted with them for generations (Agrawal 2006). And sometimes it is stopped only because richer, more powerful interests want to do the deforesting themselves, as in Thailand, where relatively harmless slash-and-burn was stopped only to throw the forests open to military-backed logging firms (Delong 2005).

Note that this includes dispossession of local people to make room for conservation. Rural and indigenous people have often been displaced or denied access to their traditional resources by national parks, tourist developments, forest conservation, reforestation projects, and the like. In East Africa, the Maasai and their neighbors maintained the land and its wildlife. When these peoples are displaced by game parks, the parks promptly deteriorate and wildlife populations decline (Fratkin 2004; West, Igoe and Brockington 2006; McCabe 1990, 2004; McCabe et al. 1992). Similar stories from all over the world are painfully common knowledge to researchers.

Particularly valuable are several recent books that address the tendency of First World environmentalists to ignore local people. Most of the anthropological literature concerns indigenous and Third World people. (Some particularly noteworthy examples of a generally high-quality literature are Agrawal 2005; Lowe 2006; Nadasdy 2004; Tsing 2005; West, Igoe and Brockington 2006; West 2006; some of these studies, and several related ones, emanate from Yale, where Michael Dove and James Scott were instrumental; Scott 1998.) However, First World rural folk have been targeted too. Even relatively rich and powerful First World groups, such as ranchers, are frequently ignored, peripheralized, and even dispossessed in the name of the environment (Hedrick 2007; Sayre 2002).

Some of the stories are harrowing. James Igoe's work with the Maasai is particularly sad (West, Igoe, and Brockington 2006; Igoe, in preparation and personal communication; I can verify Igoe's findings from my own research there, on which I draw for what follows). The Maasai have been widely displaced from their grazing grounds in Kenya and Tanzania for "conservation" and "game protection." Ironically, the result has been widespread crashes in wild animal populations. The Maasai do not hunt; they have a taboo against eating anything they do not raise. They do, however, burn the land and then graze it. Grazing pressure is light because the herds are kept moving in search of the best grazing conditions. This burning and grazing open the land, eliminating brush and insect pests and keeping the grass in prime condition. Grass is adapted to grazing and needs it to stay healthy. Conservation by lockout of Maasai has led to undergrazing and to brush invasion. It also leads to fencing and restriction of access, which eliminates animals that need to migrate. So not only the Maasai but also the animals and the conservationists all lose—a perfect lose-lose situation. The Maasai have been allowed to return to some areas, often rather *sub rosa*.

This is an easy case. Clearly, the Maasai should not only be allowed to return to their lands, but their incredible expertise in range management should be thoroughly studied and used in cooperation with them. They should manage their lands. This is unlikely to happen in current political climates, but it is eminently possible in future.

Something similar has happened in Australia, where displacement of Australian Aboriginals from national parks led to rapid decline of wildlife and vegetation. In many cases, the Aboriginals have been allowed to return, and varying degrees of comanagement have evolved; there is need to continue and improve this relationship. At Uluru ("Ayers Rock"), for instance, local groups are now burning small patches and otherwise restoring traditional management (Anderson, in preparation, based on brief field research in 2006; Kohen 1995; Evelyn Pinkerton, in preparation and personal communication; Pyne 1991; Deborah Rose, 2000, 2005, and personal communication).

The indigenous people I know best, the Maya of Quintana Roo, have had a somewhat more happy outcome. They are intensive, expert managers of the land (Anderson 2003, 2005; Anderson and Medina Tzuc 2005). They have preserved the forest for centuries, in spite of clear overshoot in the period from 700 to 900 that may have contributed to the famous "Maya collapse" (Demarest 2004; Diamond 2005; Gill 2000; Webster 2002). The Maya still have control of much of their lands in Quintana Roo and practice fairly traditional agriculture. They are integrating into modern sustainable-yield forestry projects with varying degrees of success (Anderson 2005; Faust et al. 2004; Primack et al. 1998), though there have been some serious cases of exclusion for "preservation," with predictably bad results. Other less than happy attempts at accommodating Maya management and modern preservation have been made in neighboring Campeche (Haenn 2005). Maya in Yucatan and in Guatemala are not so fortunate, but at least have some land still. Again, the success of many projects in Quintana Roo proves that the best way to manage Maya lands is through comanagement, leaving local communities in control and their traditional strategies intact but assisting them with some aspects in which modern biology is useful.

Other cases are more vexed. Extremely heavy-handed international efforts to conserve primates in Madagascar displaced Tanala farmers in Ranomafana, with disastrous consequences for their health and welfare (Harper 2000; her story was fully confirmed by a research team, including the present author, that visited the area in 2005). The situation is more complex, however, since the Tanala have also been displaced and sometimes exploited in their old and new quarters by their stronger neighbors the Betsileo (Anderson ms 2; see also the superb thesis of Pollini, 2007, on a comparable nearby case). The Tanala, like the Maya, are competent and knowledgeable swiddeners but are less careful with fire and more populous on the ground and seriously degrade the forest in the Ranomafana area. They also face massive invasion by introduced weedy plants that harm the environment both for them and for the critically endangered lemurs. Finding a best-case solution in this situation is complex. No one has found it so far.

Worst of all is the situation in which small-scale local communities really are depleting a world-class resource. This happens rarely with indigenous, traditional systems, which have by definition worked for a long time, but it happens more commonly in today's world. Even the extremely conservation-oriented Maya of Yucatan are overhunting game to the point of totally exterminating it, thus losing a major protein resource for the rural poor (Anderson and Medina Tzuc 2005). Government regulation of hunting is absolutely necessary to the survival of many rural people but is not happening widely. In this case, social justice would require government action restrictive of local use. Even the best of local users can abuse resources and cause environmental injustice.

The Matsiguenga, noted above, have a large and growing population within Peruvian national parks and reserves, hence John Terborgh's concern (Terborgh 1999; I have heard anecdotally that the situation has improved). The worst case I have heard of was reported by Cynthia Fowler (2007). She has been studying the Cat Tien Biosphere Reserve in Vietnam. It is being poached to death by local minority groups, desperately poor and desperate for cash. They can sell essentially anything in the reserve to the Chinese (China is nearby), who use any and all animals and many or most of the plants for medicine or food.

Conservation in cases such as the Maasai is often merely misguided (see Latour 2004). Sometimes, however, it is actually intended to shore up governmental or elite power rather than to conserve resources (Agrawal 2005; West, Igoe and Brockington 2006; compare Scott 1998). Sometimes, simple mistakes, based on the assumption that local people could not know what they were doing, have caused colonial and postcolonial powers to scorn local practices and promote alien practices that turned out to be less well adapted. This has happened especially often in Africa, where racism led to a particularly high level of indifference to local competence. The classic study by Fairhead and Leach (1996) has been followed by a large literature (e.g., Zimmerer and Bassett 2003). India is another locus classicus, and here some success stories followed from reversing such colonial policies (Agrawal 2006).

Indigenous people's rights to their land have only recently been acknowledged in many countries and are still imperfectly recognized in most. Rights over traditionally used resources are even more contested. Often, colonial governments took full title to the lands they had conquered or claimed. This land is now either government land or has been privatized. Usually, such privatization means that the land is stripped from indigenous people and given to alien colonists, as in Australia (Attwood and Marks 1999) and the Americas in the days of settlement. However, even when land is privatized to indigenous individuals or families, privatization is dangerous. The old game of "getting the Indian drunk and then getting him to sign his land away" is, amazingly, still played in the 21st century. Dombrowski (2002) provides a particularly thorough and harrowing description (see Stern 1965 for a slightly earlier and even more cynical case). Even if the individuals are not promptly tricked out of their land, the wider society and its laws generally make sure that the indigenous groups cannot maintain the community-based systems that maintained sustainable resource use.

Indigenous peoples must engage in long legal procedures to get rights to use resources. Often they cannot even pick berries or gather firewood on what was once their land. Management problems, alluded to above, become serious. Canada and Australia have seen particularly sustained and knotty problems with this issue. India in colonial times suffered similarly, and decolonialization has not always fixed the problem (Agrawal 2005, 2006).

Local rural people are remarkably enlightened in their analysis of the situation. Schelhas and Pfeffer (2008) found that farmers in Costa Rica and Honduras were extremely articulate about the benefits of forest conservation, their own problems with being excluded from national parks, and the needs to accommodate. They had many "mediating" suggestions: ways to maximize their own rights and livelihoods while still conserving forests. Similar research in New Guinea by Stuart Kirsch (2006) and Paige West (2006) again discloses extremely articulate, thoughtful, intelligent rural people who are quite aware of the needs for conservation and are trying to find ways to accommodate to it—or, in Kirsch's case, to impose it on a mine that is wrecking their lives. One would wish that the world environmental community could get better access to works like this.

Global fora that seek to take into account the "interests" of "various stakeholders" are subject to the common criticism that they reduce traditional owners to mere stakeholders—one interest group among many. Champions of worldwide political-economic developments, whether Marxists or modernization theorists, generally prefer to disregard such rights as much as possible and plan comprehensively. For the usual reasons, this generally leads to dispossession of local small communities and indigenous peoples, especially when they are seen as "backward" or "underdeveloped" and therefore a drain or roadblock on the economic system. This leads to a new political alliance: Champions of private property and of indigenous rights come together to agree that original owners, or traditional rights-holders of whatever kind, have property rights that should be respected.

The general case here is particularly serious in that it shows a blindness among environmentalists. The previously mentioned economic points about "downstream effects" and passing on costs as "externalities" have been the focus of environmental economics for almost a century. Yet here we have many environmentalists refusing to face up to the fact that they—we—are taking the benefits of Third World forest and wildlife conservation, while passing on the costs to those least able to pay: the impoverished rural citizens whose lands and livelihoods are sacrificed in the name of preservation. Obviously, if a firm is required to internalize its costs by cleaning up its pollution, *the First World and the environmental movement must internalize their costs by playing fair with these local people.* This would involve not only rent payment, which is an absolute minimum, but also provision of sources of livelihood—and not just swabbing toilets in international "ecotourism" hotels! Even the British squires who fought poaching had the common sense to hire the poachers as gamekeepers.

Many conservation organizations have now recognized this principle and are working toward such goals. From marketing shade-grown coffee (Jaffee 2007) to developing rural schools, the more enlightened NGOs have pitched in. But the First World has refused to take the overall responsibility and make even the first steps.

LOSS OF RIGHTS TO KNOWLEDGE

In this contemporary world of patents and copyrights, indigenous peoples have been fighting to get their rights to their traditional knowledge recognized (Brown 2003; Laird 2002; Vogel 1994, 2000; on the related question of patenting genes, see Stix 2006; and Rosenthal's 2006 discussion of informed consent addresses many of these problems also, in a particularly intelligent article). Individuals and corporations have often taken traditional medicines and other discoveries, capitalized them, and made fortunes from them. Perhaps the most striking recent case is the birth control pill, a Mexican indigenous discovery that has made billions for drug companies but not one cent for its actual creators. Mexican native peoples had been using wild yams, *Dioscorea*, for uncounted centuries as a fertility-modifying medication; drug firms merely purified the key steroid ingredients. Of course one could say that such crops as wheat and rice were "stolen" the same way, but they were borrowed worldwide long before the days of patenting and profiteering.

Conflicts over indigenous intellectual property rights have virtually shut down the long-standing practice of drawing on such indigenous and traditional cultures for crops, medicines, and industrial creations (Rosenthal 2006; I am also drawing on personal experience and conversations with experts around the world). This is a devastating development. It is a pure "lose-lose" situation. The indigenous people lose all the financial benefits they could get; the agribusiness, drug, and other firms lose new products and profits; and, worst of all, the world's people lose the medicines and foods. Promising cures for cancer, AIDS, and malaria, to say nothing of such less serious conditions as skin fungus, rashes, and stomach upsets, now languish unresearched. I know some Maya cures for rashes and canker sores that are better than anything in the drug store—but my lips are sealed.

Traditional peoples should clearly have some sort of rights over their medicinal herbs and other resources. However, this is not without its down side. It might not be an adequate brake on the giant drug firms, who have many patent lawyers. Heller and Eisenberg (1998) have pointed out that drug patents are now a "tragedy of the anticommons . . . in which people underuse scarce resources because too many owners can block each other" (Heller and Eisenberg 1998:698). The many, often contending, people in the traditional society that provided the drug could all block use of it—so could drug firms that developed it. Under some extreme proposals, medicine or food users would have to pay all the Chinese in the world for use of ephedra (a traditional Chinese medicine) and all Middle Easterners for use of wheat. However, patents are necessary for public or private enterprise, and this problem will be solved only through patent rules and market structures. Governments will have to protect indigenous rights—and that means the rights of the local community, not of the nation-state alone. Crop genetic rights are similarly a vexed issue (Cleveland and Murray 1997 includes some sharp debates). The whole problem would require another book. Suffice it to say that some accommodation is sorely needed to protect the health and food supply of the world while protecting the rights of both traditional and industrial producers.

Politics concerns questions of rights to property, tangible and otherwise. Who, for instance, has the rights to newly discovered, useful plants and animals? This problem is especially acute when the plant or animal is not really newly discovered but was known to and used by local indigenous people. The issue is simple in concept but difficult in resolution; whole books have been devoted to it, and the end is not yet. Logically, and morally, one might think that the actual discoverers and users of the plant or animal should have the rights. However, challenges come from two sources. First, the national government of the country in question may demand rights—in spite of the fact that the national government in question may well have been practicing outright genocide on the people in question. Brazil, for instance, has demanded rights to useful plants (see "Brazil Wants Cut of Its Biological Diversity," Elizabeth Pennisi, *Science* 279:1445, 1998), including those discovered from Native American healing practice, though Brazil has a long record of exterminating its Indians. It is hard to see how governments that regularly oppress and disenfranchise, and frequently murder, their minorities have any moral rights to minorities' knowledge.

Drug firms or agricultural development firms that have to test the plants or animals for world markets have to consider their cut. It is exceedingly expensive to develop a new drug, food, or industrial plant product, and most such developments fail on the market. The companies simply have to get a high return on their few successes to make up for all the failures; otherwise they simply cannot test and develop products. The low overall return rate is one of the factors that has led to consolidation of drug firms; there are now very few of these in the world, and they are very large indeed. Moreover, they naturally tend to take a conservative position on developing new drugs. They have to be very well convinced of a drug's potential before they will bother with it.

In so far as money paid to indigenous peoples, or local governments, come out of their profits, there is that much more disincentive to develop the drugs. Many local governments seem not to understand this and are apt to demand high cuts of drug profits. On the other hand, the drug firms are not in business for their health (in spite of what they sell) and have not always been eager to deal with the question or to guarantee everyone a fair return (cf. Shiva 1997).

The same is true in agriculture. A *reductio ad absurdum* occurred when an American firm tried to trademark the term "basmati" and reserve it exclusively for the firm's own hybrid rice, which was not true basmati and did not taste like it. Basmati is a group of very ancient and intensely flavorful rice varieties confined to the Punjab and neighboring areas in India and Pakistan. Various hybrids of basmati with other long-grain rices are grown in Texas and elsewhere. They do not resemble basmati in any of its valued characteristics, though they are good enough as long-grain rices go. (See "Basmati Battle," *Scientific American*, June 1998, p. 18; I have added my own observations.)

This theft of seeds, drugs, knowledge, and tradition has been described in Vandana Shiva's book *Biopiracy: The Plunder of Nature and Knowledge* (1997). Shiva provides a long list of specific cases, mostly involving large agribusiness firms—but also involving public agencies, for governments have also been known to play the pirate. On the other hand, it is sometimes too easy to accuse firms of biopiracy when they acted in good faith but with imperfect information; the problem is currently under heated debate.

We can get a great deal of guidance from the vicissitudes of this movement (Anderson ms 1; Shiva 1997; Vogel 1994, 2000). Like the case of minorities afflicted with pollution in the United States, it represents a situation in which small, often rural minorities are subjected to injustice because of environmental choices.

In one interesting case, the Tulalip Tribes of Washington (2007; what follows is based largely on information from several conversations with Preston Hardison, the tribe's consultant on these issues) have drafted a cultural protection act that is intended to give the tribes full powers to copyright their intellectual property under their traditional copyright laws. Unlike most peoples of the world, the indigenous peoples of the Northwest Coast had a full system of copyright law long before European contact. Songs, traditions, tales, artistic motifs, and other cultural creations were owned by individuals or corporate descent groups, ranging from families up to very large but usually kin-defined groupings. The heirs of these corporate entities are the modern tribes, recognized by the United States as "dependent nations" and as corporate entities. Thus they can exert their right to set such policies. The intent is to allow them to get the full benefits of conserving their subsistence resources and management systems, getting full returns for use of their art motifs and tales, and so on. This will stop such things as the blatant appropriation of native art motifs for advertising and other commercial purposes, which is not only offensive (especially if the motifs are religious symbols, as they often are) but makes a lot of money for the offenders—money that could go to the actual indigenous owners of the motifs.

Similarly, Will McClatchy (personal communication) and his group at the University of Hawaii have developed model protocols for working out agreements with local governments—national or community—to divide up the rights and benefits of bioprospecting.

Unfortunately, these models cannot always be copied elsewhere. The Yucatec Maya of Mexico, with whom I am familiar, are in a far more common situation. They have no corporate tribal identity. Their descent groups are diffuse and not legally recognized. They blend imperceptibly into the "mestizo" population (very often, individuals call themselves "Maya" in one context and "mestizo" in another). Their traditions are mostly shared by neighboring indigenous and nonindigenous peoples. There is no way they can copyright anything under present laws. Individual healers reserve rights to their special knowledge, and *ejidos*—corporate landholding communities—exercise rights as they can, but anything widely shared by the Yucatec has long escaped into public domain. It is hard to imagine how this situation can be remedied.

Of course, this is merely a special case of the general fight over intellectual property rights, which waxes ever hotter as China disregards copyright laws, college students download music illegally, and "hi-tech" vies with "big pharma" in court. (High tech industries want looser rules, drug companies want tighter ones; see Samardzija 2007.)

Indigenous, minority, and rural strategies for managing resources have been devalued and depreciated, in spite of the fact that they are often known to be superior to any others available (see Schelhas 2002).

BALANCING ACT

The worst problem for philosophers of environmental justice may be balancing the common good against local rights. If all humanity suffers major damage due to resource misuse or overcontrol in a local area, normal moral considerations would privilege the wider interest. Interests of a nation are less clearly privileged, but are not inconsiderable. Often, in environmental conflicts, biologists privilege common humanity, sometimes to the neglect of local interests; anthropologists and many nationalists privilege local interests, often to the detriment of humanity at large. Everyone recognizes that individuals and local communities often have to suffer for the greater good. They have to pay taxes, defend the country from invasion, endure quarantine in epidemics, and much else. The question, then, is when or under what circumstances this becomes unjust.

Other fairness problems seem even further from the economic nexus.

Consider, for instance, the obligations of the scientific community. Scientists fiercely guard their independence to study what they want to study and to follow the most exciting new leads rather than the most socially useful ones. Research funding has gone into high-energy physics, molecular biology, and other areas, exciting enough, but unfortunately displacing organismal biology, soil science, research on traditional ecological knowledge, and other directly environment-related studies. Organismal biology and social science have suffered in particular.

Another difficult concern is intergenerational fairness. This is much more difficult to analyze because we do not know what future generations might need or want (as in so many cases, the best discussion of this issue to date is in Shrader-Frechette 2002). A prudent conservationist in the 1880s might have tied up vast tracts of Europe and America by designating them for oat growing to feed the horses so necessary to transportation and thus made it impossible to build the highways and airstrips that soon became the necessities as automobiles and planes (both barely imaginable in the 1880s) displaced horses. Conversely, 19th-century hunters exterminated many species, never thinking that future people might actually want to use these animals or simply have the pleasure of seeing them. Probably the major moral charges here are the classic line "leave some for others" and the wider sense that, whatever else is true, we have no right to eliminate completely the *possibility* of future generations having the choice to use or enjoy a resource. In other words, exterminating species and completely exhausting specific resources are absolutely incompatible with intergenerational fairness. The precautionary principle has been usefully applied in this regard (Cooney and Dickson 2005). We can also be reasonably confident that future generations will not want vast nuclear waste dumps on usable land.

Serious also is cultural labeling of certain groups as deserving of less protection or consideration. In the United States, less affluent black communities and Indian reservations are notoriously targeted when anyone wants to locate a polluting industry or a toxic waste dump (Bullard 2000). Other countries have equivalent populations: "others" whose communities are more or less fair game (e.g., Roma "Gypsies" in Slovakia; Barbara Anderson, unpublished research). One major cause is the "not in my backyard" (NIMBY) syndrome.

Big dams that displace poor communities are justified through general-good rhetoric, in spite of facts. Thayer Scudder, an anthropologist who is the world's expert on resettlement of people displaced by large dam projects, has recently written a scathing report on this issue and on big dams in general (Scudder 2005). In spite of his hopes that big dams can be beneficial, he has not found a project in which they clearly are. (Neither have several other experts I have consulted; they might not like to see their names here.) Big dams cause vast environmental damage, and in the Third World they usually displace large numbers of impoverished people who never have a chance to rebuild their lives (most of Scudder's book concerns this issue). Moreover, dam projects always incur cost overruns, usually massive ones. Cost-benefit accounting is routinely "cooked" to cover up the resulting diseconomies. (The same is done for countless other ill-conceived development plans. I have personally seen literally dozens of cases.) Correcting this distortion makes dams look most unjust and unbeneficial, especially if we assess the actual damage to the lives and livelihoods of the poor.

One tactic used by large-scale polluting and overexploiting interests is the "jobs versus environment" rhetoric. Controls on such activities as logging are said to be costing jobs. So, for example, saving spotted owls by protecting forests costs loggers' jobs. Foster (1998) and Goodstein (1999) provide solid, no-nonsense accounts of this archetypal controversy and argue successfully that environmentalism fails without class awareness—in this case, sensitivity to the real job loss problem that sustainability would bring. The countless logging ghost towns all over the American West remind us that when all the trees are cut, there are no jobs. Sustainable logging, fishing, and other extractive activities maintains fewer jobs initially than cut-and-run strategies but, of course, maintains many more jobs over the long run (Goodstein 1999). This is the most obvious form of short-term versus long-term strategies. Yet the jobs-versus-environment rhetoric continues to win adherents, even among groups otherwise concerned with environmental justice. This cuts in both directions: Environmentalists can be incredibly insensitive to workers' genuine needs (West 1995).

Selective risk perception is involved here. People in general are overoptimistic, so may miss warning signs or be overly hopeful of their ability to cope (Anderson 1996; Taylor 1989). In the United States at least, studies indicate that white males see less risk than other groups, and where they are dominant, caretaking may suffer (Satterfield 2000; Satterfield et al. 2004).

Individual rights rarely seem to count for much in these situations; communitarian morality triumphs. In fact, in matters of environmental justice, communitarian ethics are very often part of the problem rather than part of the solution. Much more serious than "greed," or indeed any rational economic interest, is racism, especially the automatic, unadmitted kind typical of nation-states dealing with small minorities. It is simply assumed that the Roma, or Maya, or Navaho, or Louisiana black communities cannot know what the sophisticated government experts know, cannot manage resources well, cannot understand what is in the national interest, and do not count for much in any case (this is mostly based on my own extensive personal experience, but see Haenn 2005 on the Maya; also Agrawal 2005, Shrader-Frechette 2002, West 2006, and many other sources cited elsewhere in the present book). Information contrary to this model tends to be dismissed as "romanticizing the noble savage."

Large-scale polluters and resource destroyers often deliberately propagate hatred of the groups they victimize, as a way of defanging sympathy and political unity. They also, of course, stir up hatred of their political enemies—that is, anyone who works for environmental justice (Helvarg 1997).

Sometimes the environmentalists reply in kind, with equally unjust biases (West 1995). In Third World countries, murder of local opponents to mining, logging, and similar activities is common and even routine. It is not unknown even in rich nations.

Many authors have pointed out that governments and other powerful people use the discourse of "development" or even of conservation and environmental protection to dispossess local people and to inflict extremely destructive projects (from big dams to exclusionary tourist resorts) on local communities. To date, the most sustained and significant analysis of environmental discourse is Bruno Latour's *Politics of Nature* (2004). It and many of its lineage suffer from a rather naïve belief that fixing discourse can fix the environmental problem. Discourse, however, is not an autonomous, abstract thing (as Latour appears to believe); the term refers to our ways of talking about reality or what we perceive of reality.

Environmental discourses have generally been complex and multivocal, with no clear or simple relationship to power and with no clear or simple "good" or "bad." One must consider ecological, economic, and political realities as well as the multivocality of real-world discourses (Scott 1998).

Much recent attention has been given to Michel Foucault's concept of "governmentality"—basically, how governments create subjects by manipulating perception and knowledge. (George Orwell said most of it better in his novel *1984,* but Foucault is more cited these days.) The word has given rise to a less pejorative one, "environmentality" (Agrawal 2005, 2006), to describe the development of environmental subjects (*sc.* "subjects" of the state, as opposed to fully recognized citizens, not subjects of research).

In all the political studies, the word that appears over and over is *accountability.* Giant firms are above the law in many countries and perilously close to that state in many others. Nondemocratic governments are, by definition, unaccountable. Technocratic bureaucracies, even in the most democratic countries, can be hidebound, nontransparent, and rigid. One need seek no further than the average state university administration to see this. Mexican Spanish has coined the wonderful words *burrocracía* and *tortuguismo* for such Weberian bureaucratic behavior. Recourse is, by definition, less and less available as one goes down the socioeconomic hierarchy. The less affluent and the more rural and indigenous pay the most costs and have the least recourse. They cannot bring the powerful to account. They cannot even obtain honest cost-benefit accounts or full disclosures of what is happening.

But the really hard cases are those such as Ranomafana and Cat Tien. Here, we know all too well what is happening, and either the world, the local people, or both will have to suffer. Continuing of the present course would be the worst option, since it will leave the local people ruined in a very short time, with the world-valued resources gone forever.

Possibly the best way to implement a better future is to find and cultivate local dedication. If the people in question can produce their own saviors, all that the wider world community needs to do is support them. A recently reported example is Dr. Pilai Poonswad of Thailand, who is saving the hornbills of south Thailand (Stone 2007; his enthusiastic account may be exaggerated, but she is clearly doing some good work). She has established good rapport with villages in the troubled south, where Thailand's 19th-century imperial takeover of three small Malay states has led to an endless, ugly rebellion that has taken religious overtones. The Malays are Muslim; the Thais are Buddhist. World Islamic radicalism has had some effect recently; there is also a comparable, though small, extremist Buddhist movement in Thailand (B. Anderson, personal communication). The Thai government continues its essentially imperialist policies in the area (Leslie Sponsel, personal communication). Yet, Pilai has developed village-level protection for hornbills, and this now transcends the war. This is, unfortunately, a rare situation because it requires an almost superhuman combination of ability and dedication.

Methods of common property management have been known and discussed for some time (Ostrom 1990; Ostrom et al. 2007), with comanagement by far the most hopeful and valuable way to accommodate traditional management with modern systems (Pinkerton and Weinstein 1995). However, comanagement is made difficult by the enormous gaps in worldview between traditional peoples and modern bureaucrats and the dubious commitment of the latter to the project (Nadasdy 2004). This gap can be bridged more easily than Nadasdy thinks, however; I have seen it done in Quintana Roo without much trouble. Nadasdy, like so many anthropologists, is a cultural idealist, who tends to essentialize culture at the expense of common humanity and to see hopeless incommensurability where there is none. Actual differences are indeed formidable, but common humanity exists, and hard work over time allows translation. After all, ethnographers do it routinely. Nadasdy himself worked with the native peoples of Canada for only a few months, yet he is confident that he knows their views. Long experience suggests to the present author that the management problem is lack of will, not lack of commensurability (see also Dichter 2003).

In any case, many recent studies have documented frequent failures of "TEK," participation, comanagement, and other programs promoted actively by anthropologists from the 1970s or 1980s onward. The problem is not that the ideas are bad, but that bureaucracies can co-opt or subvert almost anything (Scott 1998; for relevant studies, see Agrawal 2005; Lowe 2006; West 2006). In these cases, we have a problem with an unspecified interest, poorly asserted and defined property rights, and ill-defined groups. Clearly, definition of group boundaries and land boundaries would go far to resolve the conflicts, but then the question of who gets to define the group becomes paramount.

Most modern conservation schemes compensate local people in some way. Usually, now, this is by giving them some comanagement rights in the reserve and/or alternative source of livelihood. Rarely have these plans worked. Comanagement rights can be easily subverted or taken away outright. Alternative sources of livelihood rarely live up to expectations.

In any case, any form of compensation may be inadequate. People who have a genuinely religious veneration for their land will not find any compensation adequate or appropriate—an argument used by Michael Taylor to show that people are not "rational" in the narrow economistic sense (Taylor 2006). One might think of how much money it would take to compensate for willful destruction of the Parthenon, or central Jerusalem, or the Vatican.

MORALS AND EXTREMES

Much of environmental justice is based on some form of utilitarian calculus: the greatest good for the greatest number over the greatest time. This is, for instance, the normal way we assess pollution—preventing harm and calculating that the benefits to industry (a small increase in profits, at best) do not outweigh the costs to victims whose health is damaged. Sometimes the pollution generates major benefits for minor costs and is allowed. Pesticides stand or fall on this calculus, for instance.

The idea that overall benefit should be maximized, even at the expense of many people, is not classic utilitarianism, which qualified benefit maximization with Jeremy Bentham's phrase "each to count as one, no one as more than one" (Sidgwick 1981/1907).

Another form of utilitarianism is prioritism (Broome 2008), which weights the calculus by giving priority to the interests of those most at risk or most impacted. Broome speaks of the fate of the poor as global warming advances. Others have been concerned with the fate of asthmatics in an air pollution crisis and of fishermen dependent on a fishery that is being destroyed by a dam. Recall also the problems of small-scale local farmers displaced by biodiversity conservation projects; their interests are often prioritized by anthropologists and other concerned scholars.

Some NGOs and anthropologists hold that the interests of small local groups must always be given total priority and that the interests of the world community must not be considered at all (Anderson, ms. 1). Social justice demands that we give consideration to helping the poor and reducing the inequalities that doom them to poverty (Rawls 1971; Sen 1992; Scanlon 1998). However, the idea that we owe *no* consideration to the public good and *all* consideration to individuals directly impacted by a project is a different problem. In terms of ethical philosophy, it is a form of Kantianism. It involves taking to extremes the Kantian idea that individual humans must always be treated as ends, never as means. Kant himself was not so extreme (Kant 2002).

Much of the philosophy behind environmental justice issues turns on the relative importance of individual, group, and species as ethical entities. Arguably, whatever good has happened in the world in the last 500 years is due in very large part to the rise of individual rights, notably civil rights and human rights. Group interests in recent centuries have largely meant religious bigotry, ethnic hatreds, and class rivalries, all of which are major causes of injustice—environmental injustice as well as other forms. In particular, the nation-state (as an institutional form) and the world's nation-states (as actual entities) have disproportionate power over resources—they can, and do, use them at the expense of individuals and of the human species.

However, many in the environmental justice community have argued that the pendulum has swung too far toward individualism (and sometimes toward speciesism), leaving group and ecosystem interests underserved. An individual-rights activist could argue that most such cases actually involve damage to individuals more than to "community" in the abstract. Persons, not towns, get cancer and emphysema. Persons, not abstract "society," lose property values and suffer from resource loss. But communities (as such) do suffer. Their livelihoods, local governance and autonomy, and local institutions are often the first losers from environmental injustice. More important, defining certain communities (however defined) as "inferior" or otherwise unworthy is critical to subjecting them to environmental injustice.

Thus, several authors (see reviews in Liu 2001 and Shrader-Frechette 2002) speak of community rights, communalistic ethics, and individual rights. Environmental justice issues usually concern whole communities that are subjected to pollution or resource alienation. These communities may be defined spatially, ethnically, or simply by being particular groups subject to special discrimination and prejudice.

One would think that communitarian ethics (see, e.g., MacIntyre 1984, 1988) would be appropriate, but communitarians like MacIntyre advocate strong, often religious, hierarchic systems that impose values from above—the Catholic Church is a frequent model. Yet, in most environmental justice cases, a small grassroots community is oppressed and subjected to environmental damage by precisely the large hierarchic systems that MacIntyre and his followers idealize. So the question is one not of individual rights, or individual versus community, but large community versus small community. The communitarian will naturally prefer the larger one, other things being equal (at least this is true of MacIntyre and many others). This may be "majority rule" but does not serve justice to the minorities. At the very least, there is a contrast between imposing the particular ethics *of* a community and imposing universal ethical standards on a conflict situation *involving* a community. Perhaps a new type of community ethic is in order. Creating such an ethic might be a major task for the future.

ETHICS AND PROTEST

The victims of environmental injustice may unite to protest. This has produced some of the most interesting and thought-provoking literature (Cole and Foster 2001; Faber 1998; Goldstein 1993, 1995; Sachs 1995; Szasz 1994). The people sometimes even win. Within my experience, action by a tiny neighborhood group led to closing of the Stringfellow waste pits in Riverside, California. This in turn was an important case adduced in the creation of the U. S. Government's Superfund for cleaning up toxic waste sites. Such political actions, or the lack of them, transform cultures. Over decades, the United States has shifted attitude in both directions—toward environmental justice and toward deregulation of pollution.

Religion offers potential, having been used sometimes to rally citizens and, more generally, to provide ethical and motivational backing (Anderson 1996; Shrader-Frechette 2002; Leslie Sponsel, Patricia Townsend, personal communication; Tucker and Grim 1994). However, religion has notoriously been a force for domination and oppression in the past, and one fears for its wide use.

One disturbing analysis suggests that environmental ruin and injustice are among the factors causing the genocidal civil wars and state failures of the late 20th century (Homer-Dixon 1999). While Homer-Dixon's analysis appears to me to be simplistic and in serious need of refinement, its findings cannot be simply dismissed, and more research on this area is needed. Of nations suffering from civil wars and genocidal acts in that period and in the early 21st century, only Serbia was not characterized by extreme poverty, environmental degradation, dense population, and authoritarian or unstable rule. Several states with the first three characteristics survived without violence, but they were, significantly, almost all democracies. States with major environmental pressures that were also unstable dictatorships almost all suffered civil wars or genocides (again, my research, being prepared for publication; but simple inspection of cases is quite enough to show the trend; consider Congo, Guatemala, Haiti, Liberia, Rwanda, Sierra Leone, Sudan, etc.).

Most of the analysis of environmental injustice has been focused on economic and political-economic issues and on grassroots activism. However, some of the most interesting research has dealt directly with political issues. Notable is William Ascher's book *Why Governments Waste Natural Resources* (1999). Ascher finds that corruption, poor policy design, instability, inefficient or incompetent institutions, poor training, and undemocratic institutions all add themselves to the basic economic factors. Scott Barrett (2003) examined international treaties, finding that those with good ways of enforcing negative sanctions on treaty breakers are the only ones that work well and conserve resources. Other books have criticized the World Bank and related institutions for their indifference to social justice in general, including environmental justice, and for their poorly thought-out policies (see, e.g., Hancock 1991; Rich 1994; Stiglitz 2003).

Amartya Sen has made the general case for the ethics of fair development and sustainable resource use. Sen's brilliant, clear, tightly argued ethical and economic works have become the gold standard in the wider arena of development and welfare economics (Sen 1973, 1984, 1992, 1997, 1999, 2001). His analyses apply fully and importantly to environmental social justice.

John Martin Gillroy (2000) argues for Kant's philosophy, based on individual autonomy and the categorical imperative (logically derived basic moral principles), as an ideal grounding for environmental justice. However, as we have seen, this may have problems. Another Kantian, John Rawls, explicitly left environmental concerns out of his theory of justice (Rawls 1971), but his idea of "justice as fairness" clearly can be applied to such issues.

Much of the environmental ethics literature could use a great deal more grounding in serious ethical philosophy; Sen, Gillroy, Shrader-Frechette, and Rawls provide major starting places. All four are, at base, Kantians, whereas almost all the ordinary, everyday, or techno-scientific arguments in the environmental justice universe are utilitarian—mostly of the "pollution and poverty damage people's lives" variety. Environmental justice thinkers must address and deal with this disjuncture. The U. S. Constitution made a start, rather *avant la lettre*, by asserting "inalienable rights" to "life, liberty and the pursuit of happiness." This would seem to intersect Kantian and utilitarian concerns, admittedly before either of those ethical traditions was crystallized. Some effort has also come from the field of virtue ethics (Sandler and Cafaro 2005), but applications of this area to environmental justice per se, as opposed to "the environment" in general, are only beginning.

The above authors try with considerable success to avoid the inevitable danger of philosophy: that it will become empty dogma, maintained without reference to reality. However, this danger is always with us.

Following Sen (1992) and Scanlon (1998), we have to ask what we as humans owe local disadvantaged persons and what we owe all human persons. We may also worry about what we owe nonhuman (or "other-than-human") persons. This is a new way of thinking for many people of European descent, but traditional and nearly universal among Native Americans. We owe nonhumans some consideration—how much is controversial.

In 1991, minority and indigenous peoples of color met in Washington, D.C., and issued a strong plea with 17 guiding principles. These are given in Meena Palaniappan et al., "Environmental Justice and Water" (2006:120–122). The first of these is "[e]nvironmental justice affirms the sacredness of mother Earth, ecological unity and the interdependence of all species, and the right to be free from ecological destruction." One could hardly wish for a better statement of overall needs. The principles continue with calls for "self-determination of all peoples" and fair land use planning, including rights of minorities to participate therein. Notably important is principle 10: "Environmental justice considers governmental acts of environmental injustice a violation of international law, the Universal Declaration on Human Rights, and the United Nations Convention on Genocide." Military and nuclear activities are particularly scrutinized. Education and wise consumer choices are the final principles.

POLITICAL ECONOMY OF JUSTICE

Environmental injustice has sometimes been blamed on capitalism, but the basic economic processes transcend modes of production. Economic logic almost automatically forces the polluters to pass on the costs to the general public. Medieval records inform us that dyers, tanners, and other workers in noisome professions had to work in certain wards of the cities—and of course those were the poorest and most shunned areas. In India and Japan, caste barriers made many polluting occupationals (butchers, tanners, sewage handlers, and the like) unclean, and as despised outcaste minorities they were subject to social problems beyond their already serious occupational hazards.

Almost everywhere, environmental injustice is worse in authoritarian systems than in democracies. There are some striking exceptions, however. On the authoritarian side are relatively well-run (though far from problem-free) Singapore and Tunisia (personal research); "enlightened despotism" sometimes works. Conversely, democracies with weak cultural traditions of environmental concern (like France until recently) or with extreme disparity between rich and poor (like the contemporary United States) display considerable environmental injustice.

Democracy allows election of lawmakers and administrators. This is generally beneficial, and certainly democracies have—overall—a better record in environmental justice than authoritarian states. However, from the point of view of social justice, democracy has two major problems. First, when democracy is strictly "majority rule," minorities get persecuted and subjected to discrimination; democracy has to be based ultimately on minority and individual rights, not on majority rule, if environmental and social justice is desired. Second, when democratic elections are overwhelmingly determined by money—when the winner is the one who can get the most campaign contributions—the economic power of the giant corporations guarantees them a powerful voice. A firm that uses its campaign contributions to elect officials will expect them to help with subsidies and legal favors. "Free market" capitalism might be better—or worse—but no truly free market exists in the world, so we shall never know.[2]

Communism has not only failed to deliver environmental justice, it has actually created perhaps the world's worst systems. All sources report problems of top-down control and of favoring polluting industries—inefficient heavy industry and factory farming, as opposed to efficient clean industry and small farms. To this is added the constant pressure on factories and farms to produce more at all costs. The pollution of East Europe is legendary (Feshbach and Friendly 1992). East Europe wound up far more polluted than the West, but with far less wealth to show for it; the contrast between closely matched East and West Germany was especially instructive. So was the devastation of the Aral Sea and other Soviet water resources (Kobori and Glantz 1998).

These are nothing compared to the recent devastation of China, where pollution, deforestation, farmland degradation, grazing land desertification, and every other environmental problem have been raised to uniquely horrific levels (Anderson, personal research and observation; Brown 1996; Economy 2005; Edmonds 1998; He 1991; Smil 1984; Bryan Tilt, work in progress).

In China, socioeconomic differentiation has reemerged after the rough equality of the Mao years. The costs of development have been overwhelmingly passed on to the poor. Moreover, China's traditional sustainable and efficient methods of agriculture and forestry were replaced (especially in the last 30 years) by waste on a scale beyond that of most of the western world. China copied the worst of Soviet and (later) American practice without the care or efficiencies. In fact, China's spectacular growth rate may be largely an accounting trick; the costs to the poor, the minorities, and the rural residents may approximate or even equal the famed benefits to the well-off (Brown 1995; my research confirms this). The Chinese, and sympathetic observers, briefly hoped that reducing inequalities in income to a very low level would reduce injustice. In the end, inequalities in power led to the return of inequalities in income—possibly greater today than in Imperial China.

The world environmental crisis today is, I believe, largely the result of environmental injustice. Giant firms and complicit governments have the power to pass on costs to the least powerful, especially minorities and indigenous peoples. The firms are forced to do so by competition and the logic of subsidization and cost externalization. So are countless small individual actors—who may also be driven into damaging the ecosystem because they have been reduced to poverty by dispossession and alienation of their livelihood bases. Desperate people take the last fish or deer, knowing they are destroying their future but unable to face seeing their children starve today. Environmental justice—equality before the law, accountability and recourse, democratic negotiation, and secure equal rights including community or individual tenure—would force efficiency and make sustainability profitable.

One common problem with recent literature in these general areas is that the concept of "power" is often too little analyzed. (There are many honorable exceptions; see review in Robbins 2004, and, for one example, Agrawal 2005.) Power often emerges as an abstracted force, a thing that has a life of its own independent of people and economics. (This usually seems to stem from a very naïve reading of Nietzsche and Foucault.) Neo-Foucaultian writers also portray "government" as a single, unified entity that talks with a single, unified discourse and set of practices (governmentality). In fact, governments are often monumentally disunited, with major conflicts among bureaus and politicians (Agrawal 2005; Ascher 1999). Environmental damage and injustice may be effects of outright paralysis from disunity rather than of malice (McCay 1998; Mosse 2006).

In fact, it is probably better to see power as a shorthand for various sorts of differential access to and control over resources. Political power, military power, economic power, status, ideological authority, and social capital all differ from each other in concept and effect, however much they can be transformed into each other or combined in individual people or agencies. Structural power is not the same thing as the individual power of a dictator. Different kinds of power may call for quite different strategies. In economics and ecology, the system is theoretically open, because a creative mind can increase total wealth by inventing a more productive, efficient process. Conversely, in politics, one person's power usually can come only at another's expense; only one person can be president. Mass movements or the rise of civil societies can sometimes empower everyone together, but this is not the rule. Power exerted through ideological or "discourse" manipulation is still a different game, not amenable to easy analysis. Like basic philosophy, this area is in desperate need of major theoretical refinement.

WHAT CAN BE DONE?

My conclusions on what might morally be done to deal with some of the above problems are as follows:

1. Initial documentation remains an immediate need. We need to identify each specific case of impact by a polluter or by resource destruction on a particular community. While the stories often seem to display a monotonous and depressing sameness, each one is uniquely and vitally important to the people actually involved—on the ground—in the community in question. Moreover, each one adds its bit to the statistical total for analysis. By contrast, the field is well supplied with abstract philosophical and programmatic statements. We need data-driven studies and generalizations to address the issues raised.

    Some areas are particularly in need of further documentation. These include alienation of resources and land, intellectual property rights issues, and food security. Probably the most active debates today are over indigenous rights to control resources versus government takeover and versus the common good. These resources include land, trees, animals, and indigenous knowledge (notably of medicines). A number of protocols have been developed, but this is an actively evolving area, hotly contested (see, again, Brown 2003). But even the effects of toxic pollution, though very well studied, need still more attention.

2. Nations and other units—governmental polities, NGOs, firms, communities, and individual consumers—need to set sustainable and efficient use as primary goal in dealing with resources. These goals must replace throughput maximization.

3. They must then seek out all methods for doing this—and that will mean, above all, seeking out local and indigenous management strategies, since these usually have hundreds or thousands of years of accumulated experience and trial-and-error testing behind them. Of course, this does not apply directly to problems of recent origin, including most of the pollution cases. Local strategies are more useful in matters of sustainability, community management, community rights, and distributive justice. They show that the more serious questions have been answered, at least at the small-scale level. Sustainability really can be achieved and can be not only combined with but actually based on social justice.

4. In regard to accountability and recourse, most obvious is the simple need for full rights to sue to block projects or get compensation for projects ill-done. Critical as a background is the need to discuss, seriously, how to accommodate individual rights and freedoms, community values, and species survival.

5. Transparency, and full rights to sue for disclosure, run a close second.
6. Almost any progress toward democracy is beneficial. However, forced and sudden change sometimes causes chaos rather than progress, in the environment perhaps even more than in civil politics.
7. Above all, the world needs to accommodate local interests and wider ones. This has come to be called "comanagement" (Pinkerton and Weinstein 1989). It is notoriously difficult, largely because often each side wishes to maximize its power or at least get its agenda to dominate. The more powerful side (the state or the giant corporations) automatically wins. Then we are right back to environmental injustice. The more powerful side simply has to have good faith, and the less powerful one has to have recourse. However, many success stories exist. Some have been examined in detail (Agrawal 2006; Pinkerton and Weinstein 1989). Others need to be studied. Almost anything can cause failure, while success tends to be both more demanding and more informative.

   Win-win solutions are ideal; lose-lose solutions are commoner in the real world. Most difficult and thus most in need of study are win-lose situations, in which a local interest simply has to give way to accommodate a wider one (as when fishermen simply must stop fishing to save the overall fishery). The problem here is to determine when the balance point is reached. If the fish resource and its sustainable level of exploitation are poorly known, we cannot easily get the fishermen to stop fishing. If the fish are not really in danger, we should not stop them. If the fish are in serious danger, should we compensate the fishermen? Determining such matters involves both science and moral choices. Neither the science nor the moral choices are well understood.

8. This done, someone has to put it all together. A major current problem, indeed a roadblock in the global process, is the tendency of researchers and advocates to specialize so much that they know only their area and can be advocates only for their special interest. Conservation biologists often advocate protecting wildlife without much thought for local people. Anthropologists often stand up for the immediate interests of local people without thinking of wide or long-term global tradeoffs and implications. Global-scope environmentalists and developers think of global economic matters without considering local ones (as in the case of large dams). Sociologists and political scientists, NGOs, and government agencies all have their special viewpoints. We need to look more widely.
9. The analysis above implies that we are spared the need for a total revolution, but it shows that environmental social injustice is, under most circumstances in the modern world, an automatic process. The less powerful become still worse off and thus still less powerful. Economic and

political reform can deal with this, but a wider cultural shell remains. We need a strong, self-conscious ethic of justice, maintained by cultural institutions of all kinds.

10. A problem often ignored is how to evaluate outcomes of actual judgments or resolutions of cases. Evaluating projects is notoriously difficult and debatable. D'Estée and Colby (2004) and Feng Liu (2001) have made major strides in this area. They give elaborate, extremely thorough, well-tested methods for evaluating in detail the many aspects of projects and decisions that affect environmental justice. *Training people in this art, and actually using it, appears vitally important and seriously neglected.* Projects must, in addition, be designed with ease and straightforwardness of evaluation in mind, as part of a wider concern with transparency (A. P. Vayda et al, successive postings to Eanth-L Listserve, April 2006).

11. Finally, *we're all in this together.* That perception, translated into ethics, morals, and policy, *is* environmental justice. Such translation is humanity's only hope for survival.

"All Politics Is Local," and
All Is Now Global

POWER CONCENTRATION

Throughput of resources is to be maximized at all costs: this is the basic philosophy of many governments and of transgovernmental agencies like the World Bank and International Monetary Fund.

This is held as a *moral* decision, usually on the "jobs versus trees" theory, often also on the "consumer preference" theory ("people want goods, not parks"). Even inefficiency is held up as a moral charge; politicians often hold that efficiency would put people out of work. All reputable economists, from conservative to liberal, hold the contrary, so far as I can ascertain, but their opinions seem neglected.

Being a moral charge, throughput maximization is upheld even when it is flagrantly diseconomic. Communist East Europe possibly represented the extreme in this, but Americans will be familiar enough with such logic. For instance, throughout the latter half of the 20th century, almost all American local governments neglected or even disinvested in public transportation, and zoned cities so that workplaces, shopping areas, and residential areas were kept far apart. Meanwhile, the federal government subsidized development of oil resources and production of gas-guzzling cars, while refusing to pass efficiency standards and virtually eliminating research into fuel economy. Throughout all this, politicians defended the pursuit of increased oil production as necessary because "the people want gas." The giant firms that lobbied for the increase identified it as properly American; readers may remember years of attacks on small, economical cars as "foreign" and "un-American."

The throughput economy is now concentrated in the hands of relatively few giant multinational firms, especially the areas of oil, mining, and logging, but also in much of agribusiness. The large oil firms have incomes that are many times as great as the gross national products of the poorest nations (Juhasz 2008). The great primary-production firms are maintained by heavy subsidies, including military protection in many field sites. They have far more political power than any local community can mobilize. Even if these primary-production firms had the best of intentions, they would be trapped in a system of competition to maximize resource production and throughput. That they do not always have the best of intentions is shown by their tactics in Third World countries, for example, the complicity of oil corporations with slave labor in Myanmar and violence and corruption in Nigeria.

However, these giant firms are far from the only offenders. Remember the lawns and cats. Remember also the countless small farmers who overuse pesticides and make a virtue of defying environmental protection. Ordinary people have bought into the virtues of unnaturalism, mass throughput, mass waste, and mass consumption. *The system, not the giant firms, is the problem.*

By contrast, traditional societies considered the future and were aware of resource limits, and thus were less prone to idealize throughput and maximize waste. They simply could not afford to. They were not necessarily more virtuous or self-sacrificing than we are. More than a few were quite irresponsible about resources, in fact. They just had different economic realities to face, and different ideological equipment to use in facing those realities. They were rarely if ever in a situation where they had the chance to maximize throughput as a virtue. Traditional elites that hunted simply to see how many animals they could kill would be one case of doing so; the rather frequent occurrence of such hunting in world history shows that we of the contemporary world economy have no monopoly on vice. We, however, have systematized it to a unique degree, and now have to stop it.

CORPORATIONS AND BUREAUCRACIES

As many have pointed out, the "resource curse"—and inevitably resource misuse and environmental problems in general—"is primarily a political and not an economic phenomenon" (Karl 2007:256) to be solved by transparency and other aspects of accountability rather than by simple economic formulas. As she puts it, "petroleum dependence turns oil states into 'honey pots'—ones to be raided by all actors, foreign and domestic, regardless of the long-term consequences" (Karl 2007:257). Everyone seems to agree that oil is always worst (Humphreys et al. 2007), but the same is true, to varying degrees, of states with mineral wealth and states based on export-oriented industrial agriculture.

Some national and international logging firms have wreaked havoc in Malaysia, Indonesia, Brazil, and dozens of other countries (see, e.g., Peluso 1992). And oil wealth has actually been well used in several states. As usual, the problem is not the firms alone, nor is it "capitalism" or "socialism." Oil corporations range from genuinely private through parastatal to totally nationalized, but they all do essentially the same things. (Under the George W. Bush administration, incidentally, oil firms such as ExxonMobil were de facto parastatals in the United States; Bush and Cheney were oilmen, and the oil companies wrote the relevant policies and were fused with government. The administration's energy policy was set in a notorious secret meeting between Cheney and corporation representatives; the minutes remain secret.)

Societies need elders and leaders. We tend to look up to them, and we follow them. This seems to have begun in tribal wars (see excellent discussion in Van Vugt et al. 2008). Status-seeking and status emulation is thus often a defensive and competitive game. Yet, we need it, to organize in the face of emergencies and conflicts (Van Vugt et al. 2008).

Bigness and complexity have their own system costs. Max Weber pointed out long ago (Weber 1946) that a large system, integrating many highly divergent specialties, requires much effort and talent to manage. We then have the managerial revolution: more and more managers, with more and more specialized skills in organizing and people-working, appear, and their political and social skills become all-important. Traditional physical and intellectual skills lose ground to administrative skills. Making one's own bread and fixing one's own house are less important now than knowing how to organize a unit or form a network. Even traditional moral virtues lose out; blunt honesty and hard physical work were idealized in my youth, but are no longer. Blunt honesty is a downright cost in a bureaucratic system. Politics requires at best indirection, and at worst outright lying.

An organization needs to raise the benefits for actual good works, and raise the level of accountability *to the system goals* for any leaders and administrators. The corruption of religion and of academic life by increasing bureaucracy was already known to Weber, whose descriptions of it are classic (Weber 1946). He saw that sclerosis from bureaucratic inertia was even more pervasive than actual abuse of power. Bureaucrats live under enormous incentives to dodge responsibility, minimize work load, pass the buck upward, and delay all change. Ultimately, a system often emerges in which a power-mad bully takes over a whole organization of foot-draggers, causing system paralysis. (Scott Adams' *Dilbert* comic strip portrays this beautifully.) More often, leaders simply want status and feelings of command, but the result may not be much better.

Corporations add to this mix the fact that they are considered to be "individuals" under law, but are unlimited in age, size, wealth, and power. They can grow to a size that guarantees them immunity to government oversight, at least in most of the world's countries; the biggest corporations (which include the major oil and agribusiness firms) each have more wealth than the 20 poorest countries combined.

All this comes to one institutional conclusion: It isn't "government" that is bad and "private enterprise" that is good, as capitalists say, or "government" that is good and "private enterprise" that is bad, as socialists say. Big government and big private monopolies are both bad. Small government and small businesses may not always be good, but at least they have the option to be; they are not driven inexorably to abuse power. Any group, and above all any institution, with total or extreme power has such an overwhelming need to maintain and consolidate that power that it cannot long resist the drive to oppress. Small, local, responsive government and firms have contrary pressures. They are near enough to their clients to find that providing service is a better route to success than providing tyranny. The local postmaster isn't the faceless bureaucracy, and the local grocery store can't be as unresponsive as Wal-Mart.

Within the environmental universe, energy corporations and international fishing interests are among the most unresponsive. We have looked at oil several times before. "The stories are legendary: In Angola . . . a billion dollars a year representing about a quarter of its oil revenues disappears . . . in Kazakhstan, President Nazarbayev has hidden more than a billion dollars in a secret fund in Switzerland . . . in Equatorial Guinea, major U.S. companies pay revenues directly into a Riggs bank account under President Obiang's direct control; and in Congo Brazzaville, Elf Aquitaine . . . financed both sides of the civil war" (Karl 2007:268).

Let us look more closely at the case of Equatorial Guinea. An attempted coup took place there in 2004. Simon Mann, the mercenary who organized the fighting, turned state's evidence, and testified that certain oil corporations backed the coup. Mann believed, also, that certain governments turned a blind eye on it. Margaret Thatcher's son was heavily involved. It is awfully difficult to imagine this happening without the backing of the whole political and economic machine that Thatcher represented (Fletcher 2008). Since then, both national leaders of Equatorial Guinea have been assassinated.

Also revealing is the Supreme Court's final disposal of the Exxon Valdez oil spill case. On June 25, 2008, the Court ruled that Exxon (now folded into ExxonMobil) was liable for direct personal damages (calculated in an extremely minimalist way), but for nothing else—no punitive damages, no wider environmental charges. The case had been in court 19 years, thanks to

Exxon's delaying tactics, and many of the people who would have received the (small) damage compensation had died in the interval. To all intents and purposes, Exxon had won—gotten off with hardly a slap on the wrist. The complete domination of extractive corporate interests over civil society was assured. Note that this was not the result of free enterprise or capitalism; it was a case of a heavily subsidized firm, closely tied to the current administration (the oilmen George W. Bush and Richard Cheney), getting special favors from the government, including the Supreme Court with its Bush appointees. Parallels with free enterprise and free competition are hard to find, but parallels to the state oil firms of Mexico, Indonesia, and China are exceedingly close. We are dealing with the unification of government and extractive industry here, not with capitalism or free markets.

One reason Exxon made such a fight of the issue was its precedent value. The Supreme Court has now set a precedent, making future punitive damages in environmental cases exceedingly unlikely. It is no secret that the far right wants to outlaw class-action suits and most suits against corporations. Without mass citizen action, they now may well get their wish.

As to fish, consider the case of Mauretania (Clark and O'Connor 2008). Reduced by poverty to trading away its rich fishery to foreign nations, it saw those nations overfish its waters into barrenness without paying much for the privilege. Worse, some foreign fishing boats hired Mauretanian fishermen (ruined by the overfishing), set them out in small boats, and often sailed away, leaving them to shift for themselves on the open sea. Such is the morality of the throughput economy. The foreign fishing boats involved are not major players; they are small operators. They can operate with near impunity, however, because Mauretania is so weak. The point, again, is that there is nothing especially immoral about big firms, still less anything especially moral about small ones; the problem is that in a throughput economy, where the government actively supports extractive industries, *only a strong and unified citizenry* can enforce checks, balances, or accountability. Ordinary people have no recourse unless they can unite.

Today, that increasingly means uniting across national boundaries. Even if all Mauretanian fishermen united, they would have little power against Korean or Russian fleets. The global power of giant oil corporations makes even the United States almost helpless before them, let alone Equatorial Guinea. The environmental movement has to be one united global effort.

A slightly different case of bigness distorting performance occurs in health care (see, e.g., Wagner 2006). This is an environmental concern because of the huge health problems caused by environmental decline. The old-time local doctor acted very differently indeed from today's health maintenance organizations (HMOs). Over 100,000 people die in the United States each year from medical error, much of it because medical staffs are over-worked and overmanaged by HMOs. Along with this have gone skyrocket-ing costs of medical care, increased inefficiency, and misallocation of labor resources. In spite of the reputed ills of "socialized medicine," no other devel-oped country has health statistics as bad as America's. In fact, care is better in much of the less developed world. This was not the case when real competi-tion existed—when doctors were private providers and HMOs did not exist.

The most critical part of the mix is accountability. Competition for con-sumer dollars is one way of forcing this, as Adam Smith pointed out. How-ever, this occurs only when the consumers are independent and able to choose. This, however, is a thing of the past for most of us. We deal with a few giant firms, not with many small producers. We talk to computerized "options" instead of humans. One cannot find a human to talk to on the cor-porate phone lines. This is not merely an inconvenience; it prevents account-ability. The firms could easily set the automated message lines to receive comments. They do not.

Legal recourse, grassroots democracy, and so on, provide alternative routes. This requires massive unity, however, to balance out the strength and the powerful economic motivation of those above the law.

Through history, people have moved upward mostly through developing their subsistence systems. Adam Smith, Karl Marx, Lewis Henry Morgan, and many others noticed this, and listed what they thought were key inven-tions—agriculture, factories, writing, and so on. Marx developed a theory that deduced modes of production from the interaction of subsistence tech-nology and control patterns.

They were perhaps too quick to foreground subsistence. They overlooked the importance of transportation, communication, security in general (including military knowledge), and social institutions. Granted, develop-ment of subsistence techniques still explains most of the changes, just as opti-mal foraging theory predicts most of the activities in the day-to-day life of a hunting and gathering group. But "most" is not "all," and what remains unexplained may yet prove to be instrumental.

Especially important is the pyramiding of power over time. Ibn Khaldun, the 14th-century Moroccan historian, saw it as driving history (Rosenthal 1958). An economy in which the rich own everything is a very different world from one in which control and wealth are evenly distributed. Ibn Khaldun saw an egalitarian society as one with generosity, loyalty, and fru-gality as virtues. Gradually, its elites become powerful. A world in which a tiny elite controls the whole system is easily corrupted, and inevitably prey to revolution—or, worse, becomes subject to the random violence of the totally alienated, who are often those who put the government in power in the first place. Then the government breaks down, and a new group takes over—ideally, one with loyalty and egalitarian values. The cycle begins again.

The collapse of the banking system in 2008, and its enormously expensive bailout by the U.S. government, is a fine case in point. Economists had long predicted that this would be the end result of deregulation and the consequent rise of a "virtual" economy, where wealth was generated by speculation, rather than by producing goods. Now the United States will pay for decades for its unwise deregulation; money will be sucked away from solving the environmental crisis, among other things.

The connection of big money and anti-environmental causes is not unrelated to campaign donations. A long and detailed survey in the *Los Angeles Times* (Sept. 21, 1997, pp. A1, A22–23) revealed much that other media hide. The situation is probably worse since—and one proof is that nobody publishes media stories like that any more. The investigative journalism that once characterized the *Los Angeles Times* and other large papers is virtually extinct.[1]

PRIMARY PRODUCTION AND THROUGHPUT

The dominant dynamic in world politics today is the rise of the giant primary production firms (see, e.g., Harich 2007). Big oil, big mining, big agribusiness, and similar interests have been growing for over 100 years, and the largest firms now dominate the world economy.

World political systems are sometimes said to be converging on democracy. They are not. Except in the still-democratic smaller nations of north and west Europe, they are converging on what seems to me is a single system, characterized by several features:

- Highly centralized government.
- Top-down control and planning.
- Direct government subsidy and support of raw material production and resource throughput.
- Subsidy of other big, politically powerful firms.
- Increasing fusion of government and business interests—either through fusion of government and business elites (as in the United States and in Mexico), or through opening up of business as government enterprise (as in China), or through corrupt influence on government by the productive sector.
- Increasing control by the resulting government-business complex over media, and thus over knowledge and beliefs. "Free speech" becomes increasingly meaningless in a world where a tiny fraction of the elite controls all

public means of communication (Bennett et al. 2003.) The Rupert Murdoch empire now includes Fox News, the *Wall Street Journal*, the the *Times* (London), and hundreds of other major venues. It is strongly anti-environment. Rupert Murdoch's corporate takeovers have led to dominance of a huge percentage of the media by his extremist right-wing views; his Fox News became quite openly the public relations arm of the Republican Party in the United States (on Murdoch and his empire, see Pooley 2007). No one has been able to challenge his right to take over the world media.

- Decreasing governmental involvement in welfare, education, health care, and other long-term investments in human and social capital.
- Resulting distortion of government: more and more, it becomes a device to consolidate the power of giant firms at the expense of everyone else.
- Resulting distortion of economics: The same giant firms become more and more involved in zero-sum political games and subsidy-seeking, less and less concerned with producing economic benefits or progress. These firms come to act in a more and more countereconomic manner: firing their best workers in massive downsizing operations, reducing efficiency, merging into unwieldy conglomerates, and the like. Everyone then rushes to copy these "successful" tactics.

The previous list is my summary of the considerable recent literature on political economy. A selection of popular accounts of this process includes Eichenwald (2000); Hancock (1991); Humphreys et al. (2007); Juhasz (2008); Klein (2007); Marcus (2002); Myers (1998); Perkins (2004); Phillips (2008); Rossi (2006); Rothkopf (2008); Sirota (2007); Stiglitz (2003); Westen (2007). I have also drawn on my personal research and on hundreds of articles and reports.

The reality is overwhelmingly one of increasing fusion between big government and big primary-production or heavy-industrial firms, rather than progressive freeing of trade, commerce, or production. Steadily dismantling health care and education is a part of this rather than a part of freeing trade; such devastation keeps labor costs down and keeps public protests from occurring. To benefit economies and help trade, the public sector that should be attacked is the military and paramilitary one, because it absorbs by far the most money and because military regimes are never friends of free trade. But the military invariably increases its wealth and power under the policies mislabeled "neoliberal." In short, we are dealing with fascism. It has evolved beyond simple fascism, however, into a truly new mode.

Autocratic states once drew their base from the military and the feudal or rentier elites. The major exception was the slave economy of the Caribbean and the American South. The genius of fascism was to see the advantages to autocrats of copying the slave economy—combining basic production and tyranny in one. This was maintained by the Big Lie, but also by the far more successful technique of dividing the working classes—setting ethnic and religious groups against each other. Hitler and Stalin perfected slightly different forms of the model. It has been widely copied since, far beyond the bounds of identifiably "fascist" and "communist" states. Many of the world's governments are now autocracies sustained by giant primary-production interests (Saudi Arabia, Iran, Myanmar, and others). Others are democracies sliding toward fascism. Still others are more mixed, but have had episodes of fascism in which the industrial-tyranny fusion has taken over, later to be driven out again; this is the recent story of Argentina, Chile, Brazil, and several other Latin American and Third World countries.

Thomas Friedman (2008) notes that in 2007 some 10 countries became more democratic, but 38 became less so, including several oil countries. He quotes Larry Diamond of Stanford: "There are 23 countries in the world that derive at least 60 percent of their exports from oil and gas, and not a single one is a real democracy." Friedman goes on to make the obvious further applications to countries like the United States.

Paul Collier (2007) has documented several mechanisms by which this all happens. (Collier, incidentally, is no political liberal.) High "rents" from resources allow wasteful spending, corruption, buying votes, buying media or at least their compliance, and shortcutting due process. This in turn leads to autocracy or to corrupted democracy. A few countries, notably Norway, escape by being mature democracies, but (though Collier does not say it) the United States is not escaping.

This is not a simple function of resource wealth, as Collier sometimes seems to imply. Oil does seem to corrupt, and the "curse of oil" has become proverbial (Humphreys et al. 2007; Juhasz 2008). Otherwise, however, the problems with resource production occur not so much in resource-rich countries as in countries that have longstanding problems with poverty and unstable or tyrannical governments, and thus little to offer *except* raw materials (Brunnschweiler and Bulte 2008). And meddling by the world's shadow-government of international trade and aid agencies is even worse (see, e.g., the later discussion of Shanda et al. 2008).

The giant multinational firms often intervene directly. Oil companies have subverted governments and, in consequence, been allowed to devastate whole regions in Ecuador, Nigeria, and elsewhere. They have colluded with the Burmese government to employ slave labor. International agricultural corporations made the phrase "banana republic" a joke, but the reality is no joke. In the early 20th century, Guatemala was economically dominated by United Fruit and its banana plantations. The Dulles family was important in that corporation. When in 1954 a socialist government threatened to make United Fruit accountable, the American Secretary of State, John Foster Dulles, directed a coup that brought in an extreme right-wing government. Many of the military at that time had been trained or influenced by Hitler-supporting German settlers in Guatemala in the 1930s. The result has been over 50 years of civil war, murderous violence, and genocide, leading to hundreds of thousands of deaths. Quite possibly, the right-wing military would have taken over even without Dulles' help, but at best the coup did not help the situation (on this case see Stoll 1993, 1999; Warren 1993, with comparisons with other cases). Many similar cases are well documented (see, e.g., Perkins 2004).

This being said, most observers have exaggerated the importance of "trade" and of such direct interventions, and have underestimated the importance of the deeper mechanisms Collier emphasizes.

This process has gone farthest in poor countries that supply resources to the world: Sudan, Indonesia, Guatemala, and similar places. It is definitely not "underdevelopment." It is the wave of the future. The changes in the United States, China, and Europe in the last 20 years make this clear.

Classically, feudal societies, as in Europe's "Old Regime," were run not by capital-owning entrepreneurs, but by landlords—those who drew their wealth directly from ownership of the land and practiced an extractive economy. The world today has almost returned to this sort of feudalism. Giant firms, like plantations of old, control primary production and use minimally paid, uneducated labor to extract it. This leads to an antipathy to education. Capitalists have to educate themselves and their children, at least in economics. The interest in learning often spills over into other areas. Landlords, in contrast, prefer an unskilled and impoverished labor force. They do not need educated workers, but they do need unquestioning docility. They do not even need much education for themselves and their children. They live by dominating, not by knowing. Traditional land-based aristocracies were noted for their duels over "honor" more often than for their wisdom. In thoroughly neofeudal regimes, like Guatemala's and El Salvador's during their times of civil war in the 1980s, teachers and educational leaders were among those who "disappeared," along with civil rights workers and labor organizers. The same appears to be happening today in Sudan and elsewhere.

Compared to genuinely capitalist regimes, landlord-dominated regimes are much less free. Landlord-based regimes are classically associated with established churches that tyrannically enforce dogma at the expense of thinking. Capitalists may grumble about the workers' rights, but they know that capitalism can maximize wealth only if the "bourgeois freedoms" are at least somewhat operational. Landlord-based regimes (including Marxian regimes—in practice) have no truck with such freedoms. They openly oppose human and civil rights. Even in the freedom-touchy United States, politicians from agrarian, landholding-based parts of the country introduce a steady stream of legislation to curtail free speech, establish Christian right-wing religion, return women to the kitchen, weaken or repeal civil rights legislation, and—in short—to restrict all the First and Fourteenth Amendment freedoms. In countries where the landlords have formed a genuine oligarchy, such as Guatemala and Peru, the situation is proportionately worse (though Peru shows recent signs of finally breaking the mold).

Landlords have much to gain by developing a hierarchy of ethnic and gender categories, and by maximizing oppression at all levels. This allows them to consolidate their position, rule the workforce by dividing it, and keep the workers unskilled. Thus it comes about that movements for civil rights and equality have generally been urban-based, typically among business and professional classes, whereas movements to "keep the whoevers in their place" have been rural and landlord-dominated.

This throughput order demands a consumer world, but one in which people make do with less and less while thinking it is more. It is a world of monocrop agriculture and monocrop human culture (cf. Scott 1998). Capitalism, in the classic sense, would have given us firms competing for market niches—and still does give us those, in places where capitalism flourishes.

Another feature of the throughput order is the immense power of the very few individuals who run the giant firms (Rossi 2006; Rothkopf 2008). The world is run by a tiny elite: heads of major states, military and political groupings, and international firms. (Rothkopf adds major religious figures and a few other charismatic individuals, but there is no evidence that they really make much difference.) The giant firms are almost all primary producers or heavy industries. There are a few retailing outfits, like Wal-Mart, but most of the exceptions to this generalization are only apparent ones; Rupert Murdoch's media conglomerate is essentially a public-relations arm of these interests, not an "information" empire.

The value of a human life falls accordingly. The U.S. Environmental Protection Agency decreased the value of a life from $7.8 million to $6.9 million in 2008 (Yahoo! News, online, July 10, 2008); the cut is more than it appears, because the $7.8 million figure was calculated in 2003, and inflation should have increased it since. One can only imagine the reasons for the decrease, especially with the huge increases in the prices of basic survival goods. It is not too difficult to do the imagining. At any rate, the George W. Bush administration seems to have concluded that Americans are worth less than they were a few years ago.

"Capitalism" has come to mean a range of things, some of them mutually exclusive. Most people now seem to use "capitalism" to describe any system in which private ownership of anything is important. Some use it to mean any regime they like, or any regime they don't like—even if it is clearly socialist by all objective measures. Originally, capitalism was the domination of the economy by capitalists: people whose resource was money, and who invested it to make a profit. Merchants were the original capitalists, but capitalism as a system did not come about until bankers and factory-owners became more important, because merchants by themselves rarely dominate a society. In a capitalist system, factory owners are people who invest their capital in getting the factory built, not the people who actually work in the factory. Employee-owned corporations are a different thing.

Excessive concern with "capitalism versus socialism" has blanked concern with the far more serious and important contrast between giant hierarchies—multinational corporations, international bodies like the WTO and World Bank, and the George W. Bush version of "international" U.S. politics—against human-scale enterprises. A private firm and a local cooperative or government bureau share key things. A giant faceless corporation and a fascist or communist state share key things. Certainly the environmental behavior of China, the USSR, and the giant U.S. corporations has been remarkably similar over the last 50 years. (Here, as elsewhere, I draw on the libertarians and what little remains of the "small government left," and I fear I will be attacked by both mainstream liberals and mainstream conservatives.)

In particular, the giant primary-producing firms and international agencies have a vested interest in opposing environmental conservation. They naturally gravitate to right-wing political parties and platforms (Krugman 2004).

Thus, for instance, rates of deforestation in Third World countries correlate closely with their dependence on international loans and the resulting "structural adjustments" demanded by the World Bank and IMF. These adjustments typically involve opening the economy to foreign investment and dismantling such "barriers" to "development" as conservation and environmental management. Only the growth of rural population has a comparable effect on deforestation (Shandra et al. 2008; urban population growth has no significant effect; see also Bunker and Ciccantell 2005). Presence of NGOs alleviates deforestation, even if they are not environmental NGOs, but democracy does not help.

The costs of having many natural resources but few governmental ones are now well known. John Fei and Gustav Ranis (1997) note that the dictatorial governments of those countries often want nothing more than to keep the countries backward. It is easier to control uneducated and impoverished masses. Moreover, landlords, used to dominating unskilled and sullen workers, are generally sympathetic to dictators and strongmen. Skilled workers and complex economies tend to produce threats of democracy, as strongmen found out from Poland to South Korea. Fei and Ranis also note that some countries are trapped at a lower stage of development because they export so many of their best and brightest workers to more developed countries, temporarily or permanently. They cite no examples, but one assumes they are thinking of places like Turkey, Algeria, and Mexico.

One may also note that entrepreneurs naturally are thickest in city-states at the centers of great trading networks: Singapore, Hong Kong, the Netherlands, Switzerland. Landlords dominate in remote areas that can do little but export raw materials: Bolivia, Paraguay, Congo. The dominance of landlords, and of the dictatorial regimes they produce, ensures that such countries stay backward. Rich, developed nations that rely on them for raw materials (and sometimes for migrant workers) are also predictably prone to do their part to ensure that backwardness is maintained.

The differential political behavior of the same people in different places is thought-provoking. The Dutch of the Netherlands are liberal (today); the Dutch Afrikaners of South Africa were the mainstay of Apartheid. Similar contrasts can easily be found anywhere that a population is split between a rich, trade-based country and a marginal, landlord-run one.

We are, thus, facing the possibility of a world of stagnation of science, technology, and trade. The "free trade" promoted by international organizations may prove to be a license for giant multinational firms (usually heavily backed or subsidized by their home governments) to become de facto rulers of much of the world. This is not free trade.

As world economies "mature," large firms and governments take more and more power in all sectors (Harich 2007). (Slave labor still is common in some countries.) Population growth is often extremely high in these countries, adding to their economic and ecological woes. Lack of health care is the major reason. Contraceptive technology is unavailable; also, with many children dying, people try to have many more, for "insurance." Poor educational access (especially for girls) is also involved. Also, religion is always a refuge for the desperate, and some of the world's largest religions have pronatalist and anti-family-planning beliefs.

Once again, I am *not* arguing that giant firms are "immoral" or that small businesses and individuals are somehow more moral and superior. I am arguing that the modern world is structurally committed to throughput, and subsidizes it excessively. The giant firms are best able to capitalize on this, and avoid the checks and balances that sometimes work for others in the system.

However, many giant firms are quite moral in their dealings, and of course many small ones are not.

Even in Big Oil, there is a range. ExxonMobil, the world's largest corporation, responsible for about 5 percent of all greenhouse gas emissions in recent history (Sirota 2008), has devoted tens of millions of dollars to spreading disinformation about global warming, but BP has been notably committed to efficiency and to searching for alternative energy sources, and Shell and Chevron have weighed in to a lesser but still significant extent on similar projects. Such ranges exist within agribusiness and other sectors. The more enlightened logging firms work for sustainable cutting, though many small internationals can be appallingly irresponsible. And of course firms not directly involved in primary production have every incentive to work for a healthy world with well-to-do consumers! Conversely, individuals, including environmentalists, are often notoriously wasteful and inefficient. Environmentalists can be amazingly good at justifying long road trips in SUVs, or repeated plane travel to international conferences. Thus, neither a "people's revolution" nor "returning power to the grassroots" will necessarily save the world, as bitter experience has shown in many countries.

My sense is that the rapid expansion of empires and settlements in the last several centuries made resources cheaper and cheaper and made planning horizons shorter and shorter. In the United States, there was always the "frontier," allowing people to overuse and waste their farms and settlements and then move on. The closing of the frontier in the 1890s did not stop this process; it merely allowed a combination of technology and filling-in-spaces to continue the frontier mentality. Similar changes were introduced by imperial powers to Latin America, Africa, and elsewhere. Cheap resources, cheap moving on, and cheap transport led to widespread ignoring of discount slopes. The future could take care of itself, and indeed, it generally did, for a long time. Alas, the reckoning is with us now. No one, or perhaps everyone, is to blame.

The rise of government ties with primary production produces a number of costly effects (Ascher 1999 remains the best account). One is that bribing or otherwise soliciting government aid rapidly becomes the only game to play. Satisfying customers becomes less and less an issue. The result is a lose-lose situation. The energy interests—oil, coal, wind, hydropower, biofuel, and so on—once might have competed in the marketplace, each one vying to be the most innovative, low-cost, and efficient. The same is true of agribusiness firms and often of logging and mining firms and of housing developers. Now, they compete in the corridors of power, seeing which one can play politics most effectively to get the most subsidies. This, de facto, becomes competitive bribery.

ECOLOGICAL COSTS OF THIS REGIME

Consider the ecological damage wrought by governmental development in the old USSR and in China (see, e.g., Feshbach and Friendly 1991; He 1991; Smil 1984, 1993). Socialism was not a great success at production, but its real failure lay in its inability to deal with the costs of extremely inefficient production.

Elsewhere, "big dams" continue to be popular. In China, the government completed the Three Gorges Dam, which all independent experts have condemned as a catastrophic mistake. When Chinese citizens raised concerns, the government simply threw them in jail and shut down their publications (Gleick 1998). The whole story is told by the leading victim of this persecution, Dai Qing, in a recent book: *The River Dragon Has Come: The Three Gorges Dam and the Fate of China's Yangtze River and its Peoples* (1998). The reasons for persisting in uneconomic construction of big dams include the political power of construction lobbies, the "face" of the governments in question, and the blind force of tradition (Hancock 1991; the disastrous effects of misguided development, noted elsewhere, are finally becoming known even to the media; see, e.g., Jukofsky 1999, Lamb 1999).

In summer of 1998, the Yangtze flooded much of central China. The *Los Angeles Times* for September 11 reported "nearly 14 million people . . . homeless, and more than 3,000 . . . killed" (p. A5). The *Times* also reported that logging in Sichuan had been stopped—banned as of September 1. A ban on most logging throughout China followed. The government finally and belatedly admitted the real problem. Rampant clearcutting in the mountains nearby was the direct cause of the flooding. Stripping the slopes there had allowed the torrential runoff. Heavy rains did the rest—but the rains were no heavier than they have often been.

China has a long history of deforestation and consequent flooding. The Yellow River has long been known as "China's Sorrow" because of its propensity to flood, but China's real sorrow was the rampant deforestation that really caused the flooding. Wetlands reclamation for agriculture and use of levees that raised the riverbed level then made the lower reaches much more susceptible to damage from the floods. Instead of learning from bitter experience, the Communist government extended the mistakes to the Yangtze.

The *Times* article quotes a petition from the flood sufferers that says it all: "For so many years, by blindly following the concept that man can conquer nature, we have built up vast, evil debts to the Yangtze River. . . . We are now swallowing the bitter fruit of nature's revenge."

Finally China heard, and banned most logging. A good deal of illegal logging continues, however, according to my friends who visit the country.

China's Imperial regimes were bad enough in their treatment of nature, but at least the ideology of the time was conservationist, and people were aware of the values and benefits of trees. The Chinese have always been deforesters; Mencius remarked on it in the 4th century BC, and Peter Goullart in the 1940s contrasted Chinese deforestation with indigenous minorities' forest protection in Yunnan (Goullart 1955; cf. Marks 1998). But at least they knew better, and they tried to maintain as much tree cover as they could, in sacred groves and protected forests. The Communists, at first, also had good intentions; they massively reforested China's denuded hills in the 1950s and early 1960s. However, the Marxist attitude that humans should struggle against, conquer, and dominate nature—a theme repeated over and over in Marx's writings—prevailed, and even the most minimal aspects of common sense were thrown out (He 1991; Smil 1984, 1993, 2004). Massive deforestation took place during the "Take Grain as Key Link" campaign in the early 1970s, and again during the period of shortsighted profit-seeking in the 1990s. It was this latter period that came to its denouement in 1998.

World levels of corruption in the primary-production sector are considerable (see, e.g., Ascher 1999; Hancock 1991). Discourses of democracy and good governance have all too often been used to hide amoral and exploitative goals (Abrahamsen 2000). Everyone exploiting Africa has been anxious to show how their pet project was just what the Africans needed to make them shape up (Abrahamsen 2000; Hancock 1991; Scudder 2005). This goes back to the days of the slave trade, which was sold as the best way to bring the Africans into "civilization" and "Christianity." Some of the arguments quoted by Abrahamsen and by Thayer Scudder are not much more subtle or moral.

Ultimately, the specter of a "worked-out" country is no longer science fiction. What happens when a country has become dependent on exporting natural resources, and has changed its government accordingly (i.e., into a corrupt authoritarian one), but has run out of resources to export? What happens when the mines are exhausted, the oil dry, the soil eroded, and the farms reduced to such small size—by rampant population growth—that there is no chance of developing export crops efficiently enough to make them competitive? This has now occurred in several small nations: Haiti, El Salvador, Rwanda, Burundi, Liberia, Jamaica (where it is partially offset by tourism). Everyone will recognize that these are not healthy states. Of course, some states with enormous natural wealth are in about equally bad shape (Sudan, the Dominican Republic, the Congo, Indonesia). Conversely, some densely populated states that have exhausted most of their resources have been stimulated thereby to become world leaders: The Low Countries, Switzerland, Singapore. So there is no simple cause-effect relationship with resources here;

the problem is caused by governments set up by, or unduly influenced by, giant extractive interests elsewhere. This is the classic problem of "peripheralization" (Wallerstein 1976).

Exhaustion is about to happen in several much larger countries, with certain destabilizing effect. Candidates include Guatemala, Honduras, Philippines, Nepal, and several African states. Most disturbing is the thought of what will happen to the oil nations when they run dry.

The decline of polities because of this is itself an ecological crisis of sorts. C. S. Holling has elaborated an ecological systems theory applying to both natural and human ecosystems. The human implications include a model very similar to the one mentioned previously, elaborated from Hollings' views by Brian Walker and David Salt. They see the declining phase as characterized by

> increases in efficiency being achieved through the removal of apparent redundancies (one-size-fits-all solutions are increasingly the order of the day);
> Subsidies being introduced are almost always to help people *not* to change (rather than *to* change);
> More "sunk costs" effects in which we put more of our effort into continuing with existing investments rather than exploring new ones. . . .
> Increasing command and control (less and less flexibility);
> more and more rules . . .
> Novelty being suppressed.
>
> (Walker and Salt 2006:85–87)

PSYCHOLOGICAL COSTS

One obvious psychological cost is excessive fixation on technology, and more generally on artificial and nonnatural things and styles. We are now so well-equipped with high-tech that we must devote a large percentage of our time to it. I am struck by the degree to which students today (especially in the sciences) have to spend so much time learning to use the equipment that they have little time for theory or scholarship.

A particularly pernicious trend is that increasing domination of political life by material throughput means that it grows steadily harder to *do* anything—harder to organize, to innovate, to act at all—but steadily easier to "chill out" with passive or virtual entertainment. In particular, it becomes more difficult to organize a meaningful social movement, because it requires so many more people and has been made so difficult to do. But even minimum exercise is difficult in a world of urbanized landscapes with high crime.

Active, enterprising nations become nations of TV, movie, and videogame addicts. Grassroots communities, formerly the real source of strength worldwide, disappear. People become more and more passive, glued to TV screens showing products and advertising, to say nothing of disinformational "news" (Bennett et al. 2007; Drahos 2002). The promised motivation and empowerment of the people by the Internet has yet to happen.

Individuals who are minions of a system—who have no control over their lives and cannot make their own decisions—lose their sense of self-efficacy, and with it their sense of responsibility for themselves and others (Bandura 1982, 1986). To function better than this, individuals depend on interactions and networks. Several recent environmentalist writers, ranging from the radical left (Carter 1999) to the hard right (P. Huber 1999), have thus advocated an individual-in-society view that is implicitly or explicitly quite opposed to both individualist and communitarian views (Carter provides some excellent discussion of this).

One effect is that enforcing laws, environmental or otherwise, becomes difficult. Command-and-control policies work badly in any case, but in conservation matters they immediately run afoul of the "Who will guard the guards?" question. If fish are to be conserved, people must refrain from taking too many fish even when the game warden isn't there; and the game warden, when present, must be reliable. One idea inspired by our belief in narrow self-interest as a universal motive is the attempt to use governmental authority—or, conversely, the "market"—to do all the work of conservation. Such reliance on one institution must fail. All-powerful institutions attract opposition rather than principled compliance. They produce enforcers who often do not believe in what they enforce.

As things get worse for ordinary people—real wages declining, environment more and more compromised—they become more passive and backward-looking. The danger of really reactionary politics is real. Fascism triumphed in Europe among losers of World War I when the Great Depression hit; subjected to both political and economic stresses, the people opted for fascism, with its promises to restore legendary glories while bringing military might. More and more governments in the world today are making similar choices. Russia, faced with absolute and relative political and economic declines in the 1990s, turned to Putin. China has moved from Communism to a governmental form almost indistinguishable from European fascism of the 1930s. Extremist Islamic governments are de facto fascist: not only based on religious bias, but also militarist, dominated by primary-production industries, and extremely repressive. Much of Africa has lost all pretense of democratic progress. The union of primary production and government has led to repressive, anti-environment governments in several Latin American countries, from Guatemala to Colombia. The United States shows an uncomfortably large number of trends in the same direction.

PUTTING IT TOGETHER: HATRED AND BIGNESS

The environmental implications of the dominance of primary-production interests are usually obvious. Less obvious is their vested interest in promoting hate. More generally, they have a huge vested interest in promoting disinformation, which they do on a large scale (see, e.g., Rossi 2006; Stauber and Rampton 1995). They can exaggerate the costs—especially the job losses—of stopping pollution and overharvest. They can maintain that pollution is inevitable if one is to have industry, and overharvest is inevitable in logging and fishing. They can make sure that research is only on immediate concerns and does not deal with long-term issues (they are sometimes in a position to suborn the educational systems of entire countries). They can maintain that the opposition to pollution or overharvest is elitist and concerned only with aesthetic matters.

Above all, the defenders of the status quo deploy their most effective tactic: destroying solidarity by raising "wedge issues." *Divide et impera*—"divide and rule"—is a time-honored tactic.

The politics of fear and hate have been harnessed by anti-environmentalists. The major effort was the work of Lee Atwater and his protege Karl Rove in the United States, linking the giant corporations with "wedge issues" of racial and gender bias. In the Communist world, the same process became even more extreme. In China, "struggles against nature" matched repressive campaigns that killed millions of people directly (see, e.g., Leys 1985). The political process is corrupted, and even the winners suffer, as they did in the old slavery days (Stedman 1988).

Primary-producer dominance goes with repression and religious and political extremism from the United States to Saudi Arabia, Iran, China, Venezuela, Sudan, and much of the rest of Africa. A list of oil, mining, and agricultural-export countries is strikingly similar to a list of the most unstable countries in the world reproduced in *Time* (July 2, 2007, pp. 16–17): Sudan, Iraq, Somalia, Zimbabwe, Chad, Ivory Coast, Democratic Republic of Congo, Afghanistan, Guinea, Central African Republic, and so on, with Myanmar, Nigeria, Iran, and others high in the lists. Even Russia and China are quite high. However, just to prove this is not the firmest of correlations, Norway—a major oil country—is the most stable country in the world according to this list, with Canada and Mexico (also major producers) not far behind. So primary production is not destiny; culture and degree of economic diversification are also major factors. Still, it remains clear that the worst thing that can happen to a country is to be rich in natural resources and poor in everything else (Bunker and Cicantell 2005; cf. Harris 2007).

Environmental deterioration surely contributes to the ethnic and religious civil wars that have raged in the last 20 years (Suliman 1999). It is probably no accident that the worst of these have occurred in countries that are not only ecologically devastated but becoming more so. Examples include Rwanda, Algeria, Afghanistan, Haiti, Guatemala, Mexico, Indonesia, Philippines, Ethiopia and Eritrea, Sudan, and others. (By contrast, the only recent example of civil meltdown in a country without major ecological crises was ex-Yugoslavia, and it had major conflicts over resources). In many of these, land conflicts are actually the root cause, with these conflicts rapidly escalating along ethnic and religious lines (see, e.g., Stoll 1999 for Guatemala; the same is true in south Mexico). A government distant from the citizens, or new and untried, is vulnerable.

In the Third World, not only was the resource base destroyed for immediate profit, but the profits were invested in repression. Indonesia's forest income was used for (among other things) weapons to "pacify" East Timor—an ultimately failed endeavor. Much of Guatemala's forests, topsoil, and biotic resources were turned into guns to eliminate the Maya Indians. Mexico's government used profits won at the environment's expense in a failed attempt to save one-party rule. The 21st century has seen these trends escalate into outright genocide in Sudan and many other countries. The Sudan government's bloody and genocidal wars on its own people are certainly the deliberate choice and open policy of its military rulers, but without support by China and other states, Sudan would not have its immunity from international sanctions. Not only has it been safe from these; it was elected twice to terms on the United Nations Human Rights Commission—apparently as a deliberate act of sabotage of the human rights agenda by certain governments, but certainly, also, as a way for certain oil-needy nations to maintain their trade. China, significantly, was elected at the same time, so that the two regimes with probably the world's worst human rights records were serving together.

Since Marx, we have known that prejudice and bias are not infrequently used by vested interests as part of a strategy to shore up their economic positions. However, there is general agreement that racial, gender, and other bias is otherwise costly to the system.

THE UNITED STATES AS CASE STUDY

The ideological and political changes in the early 21st century in the United States form a solid test case. (All that follows is closely paralleled by Westen 2007, but he does not see the economic roots; see Johnston 2003, 2007, and other sources cited elsewhere in this chapter.)

Conservatism has changed profoundly (see, e.g., Frank 2004; Johnston 2003, 2007; Kuttner 2008; Sirota 2007). The silk-hatted millionaire has been replaced by the talk show host on the one hand, and on the other by the faceless CEOs of the giant multinational corporations. Conservatism was once about elitism: European nobility, American old money. It had a major intellectual presence. Increasingly through the 20th century it has succeeded by appealing to broad-based hatred of minorities, women, gays. The old-fashioned millionaire, flaunting his silk hat while oppressing the workers, at least had some community spirit; even if only in death, he would endow libraries and museums. Today's CEOs rarely have any such spirit. There are many shining exceptions—the most famous is Bill Gates, and one thinks also of Jim Senegal (of Costco), Warren Buffett, and others—but they stand out precisely because they are rarities.

Anti-environmentalism, often couched as opposition to excessive governmental regulation, has sold well (see, for example, Barbour 1996), but racism also sells extremely well, and gave victory to many anti-environmental candidates. This often happens even in districts in which the environment should be a major issue.

The rational interests of even the giant firms would be served by long-range planning, sustainable management of things like forests and topsoil, and investment in at least minimal social services. They know it, and some are trying to get out of the trap they have put themselves in. But they are now increasingly hostage to their own creations or allies of convenience: military adventurists, corrupt politicians campaigning on pure-hate planks, religious fanatics, anti-woman fanatics, racist fanatics, anti-immigrant fanatics, and indeed a whole host of groups far beyond the pale of rational interest (by *any* definition; see Rich 2006).

Al Gore has said some of this in his book *The Assault on Reason* (2007). He stops just short of saying that the Bush administration is simply a corporate takeover of the United States, accomplished through mobilizing hatred. He is making that case, but pulls his punches somewhat: "it seems at times as if the Bush-Cheney administration is wholly owned by the coal, oil, utility and mining companies" (p. 202). Given Bush's and Cheney's backgrounds in the oil industry, this is hardly a "seems at times" matter. The old Republican support for education and conservation has been dropped. Republican worries about overpopulation have been replaced by opposition to birth control.

The most important development was the triumph of Christian fundamentalism over traditional conservatism (see previously mentioned sources, also Dionne 2008, Domke and Coe 2007). Conservatives once defended the Constitution, and thus religious freedom. By 2006, Republican candidates were campaigning on platforms that condemned voting for non-Christians.

It was condemned as "sin," for example, by Kathleen Harris, failed Republican candidate for Senate in Florida in 2006. Fundamentalism had become *de facto* established as the official religion of the George W. Bush administration. Attacks on teaching evolution consumed a good deal of politicians' time.

One major environmental problem with the use of conservative religiosity has been the destruction of population control as an agenda. Education also became a battleground (cf. Louv 2005). The primary-production economy opposed it, whereas the rest of the public felt a need for more and more education. As elsewhere, the right found common cause with anti-intellectual and conservative religious elements. The giant firms push the anti-science agenda fervently, to maintain doubts about the health problems of smoking, the reality of global warming, the dangers of poor food quality, and other issues (Michaels 2008).

The South shifted from solidly Democrat to solidly Republican. This led to a Southernization of the United States: fundamentalist religion, opposition to civil rights, opposition to regulation of guns, and some other causes spread widely. The same regional shift made the Democrats more pro-environment. Democrats from some southern and western districts are often as anti-environmentalist as the average Republican. However, there are now very few Democrats from these areas. Most Democrats represent urban areas that depend on clean water and air for survival, and on a steady flow of resources (as opposed to boom-and-crash cycles) for income. Several anti-environmental Democrats switched to the other party after the Republican landslide of 1994, and environmentalist Republicans have shifted in the other direction since. This consolidates the difference between the parties.

However, the same process led many Democrats to lose touch with the very real concerns of the primary producers. The successful campaigns of the New Deal, which reversed soil erosion, deforestation, and wildlife slaughter, have given way to considerably less successful programs, as the party loses touch with farmers and ranchers. The Democrats, in general, have lost their solid commitment to working-class issues and became a party of the information economy, the forward-technology sector of enterprise, and the educated urban class. This has led to a good deal of elite preservationist and not-in-my-backyard rhetoric, which alienated not only the working classes but many old-line liberals (cf. West 1995).

The self-destructive independence of liberals has been the anti-environmentalists' ace in the hole. The Peace and Freedom party (on the left) split the liberal vote and guaranteed the election of Richard Nixon in 1968. Ralph Nader destroyed Al Gore and elected George W. Bush in 2000. Identity politics led to the Democrats' undoing in 2008. One could point to many similar cases, worldwide, of a fractionated left losing to a unified right. The most basic difference between liberal and conservative is between individualism and loyalty. The latter almost always wins; given time, two people united can beat any number of isolated individuals!

Meanwhile, America saw the decline into virtual disappearance of several wistfully hopeful, idealistic movements: humanistic psychology, the New Age, new religions and newly-borrowed religions (Zen), environmental idealism, the women's and men's movements, the more idealistic side of the counterculture movement, and more. These movements were probably too good for this world, but they did have wide effect. They withered when American life turned ugly in the late 1980s and early 1990s.

Ted Steinberg, in *Acts of God: The Unnatural History of Natural Disaster in America* (2006), shows in detail that the "natural" disasters that plague American history were the result of outrageously bad policy. Putting a town in a major floodplain is bad enough, but rebuilding it there, at taxpayers' expense, after every flood, is truly appalling. Several towns along the Ohio and Mississippi have been rebuilt four or five times, with government money. President Clinton in the 1990s tried to make such money conditional on the towns rebuilding on higher ground, but had only some success. And we have learned too much in recent years of the vulnerability of New Orleans to hurricanes. The effects of Katrina were predicted after the last devastating hurricane, in 1968, and were the subject of a simulation exercise the year before Katrina hit. Nothing was ever done, and New Orleans was devastated by a completely predicted catastrophe for which all necessary plans had been made—but never implemented, thanks to the George W. Bush presidency and its inadequate emergency preparedness.

Big dams have not only proved uneconomic in themselves, but they have also led to a false sense of security among downstream builders. It is now not uncommon to build housing projects in river bottoms in California, and to rebuild them when they are destroyed by floods. In 1998, a huge subdivision was built in the Fresno River bottom—just after the 1997 El Niño floods had filled this usually dry channel to a depth of several feet (*Los Angeles Times,* "Flooding Doesn't Slow Plans to Build on Riverbed," Mark Arax, Jan. 13, 1997, pp. 1, 16).

THUS, BIG OR SMALL GOVERNMENT?

It will be seen that economics and morality, as well as politics in the narrow sense, are involved in these cases. Environmentalist political writings have covered the whole political spectrum. There are anarchist environmentalists like Murray Bookchin (1982; Light 1998). There are also right-wing environmentalists (P. Huber 1999), following in the tradition of Theodore Roosevelt and Gifford Pinchot. Yet, most conservatives and most Marxists agree on one thing: opposition to environmentalism. Clearly, conservation politics needs to take a new turn. Thus, recent authors such as Alan Carter (1999) have advocated new approaches that avoid the pitfalls of anarchism, Marxism, conservatism, and the other "isms." (Carter's proposal for "inter-relationism" is strikingly close to my own philosophy developed here, but Carter is a self-described "radical," whereas I have a strong skepticism about the entire radical/moderate/conservative spectrum.)

One possibility is a minimalist government—not one that is totally de-fanged, as suggested by the eco-anarchists and libertarians, but one that is dedicated to protecting its citizens and their environments but that leaves much management decision-making to individuals and networks. The higher levels of government should still set protective policy and organize the enterprise, and also set goals that can be set only at the highest levels, such as coping with world climate change or migratory animals.

The exact scale of government activities remains to be determined. Much more secure and important is the matter of rights. Interactive practice theory also leads inevitably to the conclusion that human rights—in a full, rich sense—are absolutely crucial and basic to all politics, including environmental politics.

Most other environmentalists seem to advocate aggressive government action. I believe they have not looked carefully enough at what that entails. Most blame the subsidy-and-big-dam world of the modern United States on "capitalism" or "big business," without realizing that the extreme of that sort of environmental management is (or was) in the USSR and China, and in the corridors of the de facto world government that is the World Bank and the International Monetary Fund. James Scott's book *Seeing like a State* (1998) should be the appropriate corrective. We are dealing with politicians here, not merchants. Similarly, I find myself unable to follow many "Greens" into anarchism, extreme socialism, or utopian plans to reform the world's spirituality. These schemes are impractical. They ask far more than organized religion (or such religion-like schemes as Communism) ever did, yet they begin with far less in the way of prophetic power and charisma. (See Barry's extended commentary and critique of these schemes; Barry 1999.) Fortu-nately, they are unnecessary. However, I also find myself unable to be quite satisfied with business-as-usual, as advocated by some mainstream environ-mentalists and defended by Martin Lewis in *Green Delusions* (1992). Busi-ness as usual is what has gotten us into this mess. We may need relatively few changes, but they are profound ones.

Through most of the 20th century, big-government liberalism has pre-vailed. Anarchist, anarcho-syndicalist, and cooperativist movements in the early part of the century died out—or were killed. The resurgence of grass-roots activism in the 1960s did not last long. By the 1980s, liberalism throughout the world could be identified with "big government." So could the conservatives, in spite of their claims. More and more management and conservation functions went to centralized, bureaucratic agencies of one sort or another: governmental, international, or NGO.

I think that putting all one's eggs in one basket is notoriously a bad strategy. Also, it is easy (and often all too correct) to brand big government as an enemy of local users, and brand regulation as a heavy-handed abuse rather than a necessary means of saving resources. This is troublesome enough in the United States, but far more troublesome in many Third World countries, where government is often quite openly the foe of small local groups and communities, and of their management of the environment. If the government whose troops were shooting at you last year now wants to take your land for a "nature reserve," you will suspect the worst, and you will probably be right (West, Igoe, and Brockington 2006). Such "reserves" tend to turn into hunting parks, ranches, and plantations for the political elites.

From Pyotr Kropotkin (1955; Robbins 2004) to James Surowiecki (2005), social scientists have noted that scholarly organizations, volunteer projects like Wikipedia (Surowiecki 2005), and some other organizations work exceedingly well. The common threads seem to be: First, there is an overall goal, usually either increasing or transmitting knowledge or actively helping others. Second, everyone can contribute, with some originality and autonomy. Third, there are correction mechanisms: peer review, the Wiki correction process, medical science invoked by medical-care organizations. Fourth, there is minimal chance to benefit from power. Anyone wanting "success," status, prestige, or raw bullying power is ill-advised to take on these tasks' substantial burdens for the abysmally low status they yield.

On the other hand, environmentalisms based on weak government are not necessarily better. If, for instance, we took Murray Bookchin's calls to ecological anarchism literally (more literally, perhaps, than he intends them; see Bookchin 1982, Light 1998), we would be back with a Tragedy of the Commons to end all tragedies of the commons. Actual elimination of government, in the present institutional climate, would unleash a Hobbesian war of each against all. So would devolving government to local units—states and counties—as advocated by the "sagebrush rebellion" in the western United States. The local units are often dominated by one or two extractive interests; the whole state of Wyoming, for instance, is essentially a two-industry state, dependent on ranching and energy extraction. (Tourism brings in about as much money, but is politically "invisible.") Naturally, the state government is rather sharply in favor of the current heavy subsidies for these activities, and sharply opposed to any controls on them. Many of its state and national legislators have been ranchers.

On the other hand, one may find thought-provoking the "free market environmentalism" of many libertarian authors (T. Anderson and Leal 1991, 1997; T. Anderson and Simmons 1993; Baden and Noonan 1998; Baden and Snow 1997; Hess 1992; Izaak 1996; Scriven 1997). Several of these authors emphasize the points noted in Chapter 2 about the success of local and informal management regimes (T. Anderson and Simmons 1993 is especially noteworthy here). However, freedom-oriented writers often forget that rights and private property exist because of government. Government has to be there first, to define and protect them. This inevitably means that private property is not totally free and clear; it exists under restrictions and conditions. Minimally, these include prohibitions against using one's property to damage someone else. Just as a person is not free to use a gun to kill his neighbor, he should not be free to release pollution that kills his neighbor. Government has to make rules accordingly.

One must recognize (sadly?) that people really do need some sort of hierarchic government, at least in the modern world. Humanity got along without formal government for two million years, but that was when we lived in face-to-face groups. It won't work now. Think of the problems of controlling global greenhouse gas emissions or migratory fish schools informally.

From those libertarians who can handle the greenhouse gas issue (by changing market terms to make inefficient fuel use uncompetitive), I differ primarily and focally in my concern for the rights of the victims. Libertarian thought takes much of its inspiration from Nietzsche, and much from "social Darwinism" (a philosophy made famous by Spencer and not really advocated by Darwin). The result has been a strong bias, among most Libertarians, toward advocating dog-eat-dog, weak-to-the-wall policies. The strong *should* win and should pass along the costs of production to the weak. Some even go so far as to say that the strong should be subsidized heavily by the government, as a reward for their strength—but such people have really stopped being Libertarians, and become ordinary Republicans. Most Libertarian environmentalists recognize that protecting individual rights means protecting people from other people at times, but most would seem to bend over backwards to permit the strong to take real and terrible power over the weak. Robert Nozick, the best-known philosopher of the movement, certainly comes in for this critique; he would create a world that not only would destroy the environment in short order, but would not be Libertarian for long, because the strong would simply set up total control. The extreme Libertarian view leads to a *reductio ad absurdum*—a world in which the biggest polluter or warlord would kill everyone else.[2]

Libertarians and anarchists, and even grassroots advocates, can do relatively little with questions such as migratory bird protection, soil erosion over a huge watershed, saving internationally valuable but locally low-valued indigenous knowledge, or organizing basic research. These seemingly disparate things have two things in common: they require very large-scale organization, and they do not provide immediate returns to any private party. Migratory birds cross many national boundaries, and are usually of benefit only very locally, often in places where it would be nearly impossible to use private compensation. Will Inuit hunters really send roast goose to all the thousands of people who have to protect geese along the migration path? Will tropical farmers whose crops are saved by insect-eating birds from the north send some of the fruit back to New England with the migrants?

Similarly, soil erosion control demands major earthworks of farmers at the headwaters, for the benefit of farmers downriver. The latter cannot know how much to pay which headwater farm. Thus a Soil Conservation Service is really necessary.

Most important of all is the preservation and generation of ecological knowledge. Knowledge—whether newly acquired through basic research or newly publicized from indigenous tradition—is simply impossible to valuate in any meaningful market. A bit of lore recorded today could become critical 5000 years from now; how do we assess the future for it?

There are some deeper problems with libertarian, grassroots, and eco-anarchist views. They rely on a wholly unwarranted faith in the isolated individual and on the local. Local groups that actually deal with the environment are indeed in a much better position to manage it than is any centralized government or firm. However, it is hard to maintain that individuals or small groups are somehow magically Good and the government somehow magically Bad. The libertarians, in particular, have a genuinely religious devotion to the "free market," an ideal institution that does not and cannot exist in the real world (any more than a perfect circle or a perfectly one-dimensional line can). When Adam Smith (1775) borrowed the phrase "Invisible Hand" from theology to describe the free market, he was having fun—he was being deliberately ironic. When the libertarians use it, they seem to believe literally that some kind of Divine Being is at work.

The libertarians and many grassroots environmentalists also have a strange, magical belief in "self-reliance." In dealing with the environment, all evidence suggests that communities and networks, and often national governments, really have to be involved. An individual, however self-reliant, is not going to protect whales or stop global warming without some cooperation from others.

On the other hand, communitarianism founders are on the reverse of the same rock. We need individuals. Many environmentalists who are not libertarian go to the other extreme, communitarianism. Yet conservation and environmentalism were created by people like Henry David Thoreau, George Perkins Marsh, John Muir, Aldo Leopold, Rachel Carson. These were Individuals with a capital *I*. They were not only willing, but eager, to take on the whole system. These men and women balanced out the individuals who used their freedom for selfish and destructive ends. More recent leaders, from Bill Mollison to Carl Pope, have also been rather individualistic. By contrast, specific opposition to environmentalism has usually come from communitarians: cultural conservatives, fundamentalist religious souls, and the like. In authoritarian countries, be they Communist, socialist, or fascist, environmentalists are "enemies of the state" and "threats to national security," and are treated accordingly.

Lone individuals can appeal to very wide and general social values. Moreover, a very loose society, with much individualism, mirrors the dynamic, flowing, changing qualities that we now know are ecological reality (T. Anderson and Leal 1991). Social engineering mirrors the older, discredited idea of tightly bounded and highly structured ecosystems with stable "climax" vegetation and countless homeostatic mechanisms.

Both individualism and communitarianism are "right"—insofar as they converge on a broadly interactionist position. The question is how to get the proper balance. A society needs to make its people think for themselves, but also have a strong sense of mutual responsibility.

Libertarian environmentalism has been sharply and intelligently critiqued by Scott Lehmann in a book titled *Privatizing Public Lands* (1996). Lehmann's book spends a good deal of time exposing "libertarianism" that is merely a thin veil over ripoff by giant corporations; this is not really libertarianism. But Lehmann also raises criticisms that do affect the libertarian position, and, inferentially, any of the individualist and grassroots environmentalisms. He argues that wide, diffuse, hard-to-quantify interests—notably aesthetic and psychological ones—are at issue, and these cannot meaningfully be traded off against hard dollars right now. Any market—read: any individualized system—would collapse into immediate dollar-hunting. Very large aesthetic, spiritual, heritage, and other less tangible interests would give way to much smaller, but more immediately rational and quantifiable, interests.

Even Peter Huber (1999), a self-styled conservative otherwise advocating free-market environmentalism, advocates much more attention to saving wilderness land.

These critiques are well taken, but a dedicated marketer would answer that you can easily assess people for their aesthetic pleasure; we do it all the time in resort hotels. Indeed, free-market environmentalists have pointed out that many American national forests now generate far more revenue from tourism and recreation than from logging—yet the Forest Service keeps managing them for logging, thus horribly damaging their scenic and recreational appeal, and losing money for all concerned—except, of course, the big timber bosses who are really calling the shots (Huber 1999). Moreover, there are millions of acres of land that fit the opposite case: They do not have enough scenic appeal to be very appealing to the Lehmanns of the world, but they do have great economic value that is now being lost because of perversely distorted markets. Rationalizing markets in both public and private goods, including esthetic goods, seems to deserve prior attention here.

Lehmann also argues that some resources have such "heritage" importance that they deserve special protective attention. Yellowstone and the bald eagle are like the Liberty Bell; they are far too important as national symbols to be on the market or subject to free enterprise.

John Terborgh (1999) argues that only national governments really save biodiversity effectively. He cites Texas—which has almost no public land, and a lot of endangered species—as an example. This is not entirely correct. Texas has, by nature, a large number of extremely local and vulnerable species—hence its high endangerment rate. My wide experience with both states suggests that California, with most of its land in public hands, has not really done much better at saving species. Texas's private landowners, under less fear of losing their land to endangered species, are less prone to destroy same. Moreover, being freer from restrictions, they have more options about preserving them. They can turn grazing land into game parks or conservation easements or environment-saving tourist ranches, for instance, whereas California public-land ranchers have to graze.

In fact, the need is to follow Terborgh in persuading governments to save land, *and* follow the Libertarians in encouraging private landowners to manage for biodiversity and sustainability, *and* follow the comanagement movement (Pinkerton 1989) in trying to get communities to manage better. In a world where landownership is a complex mosaic of public, private, and communal—as it is in most polities—we simply have to do all those things.

In the end, I personally would increase the government role—at all levels—in protecting the environment. My observations of Scandinavia tend to lead me to believe that their strategy of moderate-sized government and an even balance of government and business is close to ideal. But I would greatly free up individuals on the ground to manage resources with less bureaucratic management. I would eliminate government aid to primary production—all subsidies, tax breaks, Corps of Engineers projects to "develop," government-business deals. The government appropriately subsidizes roads, mail, and museums, but not environmental destruction. I would balance this by vastly increasing funds for environmental protection, education, and basic research. Money could still flow to ranchers, but for integrated land use, not for uneconomic overgrazing by cattle.

HUMAN RIGHTS AND SOLIDARITY

Human rights must be expanded to include the right to a decent environment—not as some sort of new "right," but as an inevitable expansion of the universally recognized (if, alas, not universally upheld) rights of every individual to life, personal safety, and security of property.

Solidarity is obviously an especially important goal. Without some sort of mutual accommodation, society cannot exist. Solidarity is the antithesis of conformity; it involves tolerating all sorts of people, so that one can work with them. Toleration here refers not to putting up with every sort of behavior, but to valuing people for what they are and what they have to offer, and figuring out how to use the latter for the common good. Some behaviors make co-work impossible, and those cannot be tolerated; human beings, though, deserve more respect than they usually get.

The immediate precedent to politics is some sense of community. Whatever "community" may be, all agree that it has declined lately. This begs the question of what it actually is. Minimally, it is a group of people who interact, feel some sort of unity, and decide on things together. Rules exist, including unwritten rules on the conduct of community affairs. Community is the arena for politics in the wide sense; politics in the narrow sense is typical of loose aggregations of people who do not share community identification. (Thus, there is a continuum from wider to narrower; really disunited polities, like ex-Yugoslavia and modern Congo, have notably narrow politics.)

Today, "communities" are apt to be overlapping and interlocked circles of people who are often physically distant from each other, but are united by professional memberships, e-mail, the Internet, jet travel, and the like. The world is now woven into a vast single fabric—truly a global community. Individual ties form a network that recedes into the distance. The old-fashioned community defined by co-residence and lifelong friendship is being replaced by communities of interest and communication. This allows politics on a worldwide scale, but decreases the salience of the local. There are many advantages and disadvantages to this change. The chief advantage is that worldwide activity and planning could potentially be made easy; the disadvantage that concerns us is that it causes serious problems for hands-on management. Management of resources has to be done on the ground, by people actually living with the resource.

The truly hopeless and overstressed do not participate in politics, or at any rate do not vote; they are too alienated (cf. Bandura 1982; Westen 2007). Those who do vote may often vote for economic prosperity and security, but, at least as often, they vote against and struggle against the groups they fear. They vote against, riot against, fight against, and pass laws against groups that they feel they can actually overcome and control. Occasionally, and more hopefully, they resist or rebel against more powerful groups. For obvious reasons, though, resistance to the powerful is more often through subtle means, such as James Scott's "arts of resistance" (Scott 1990). The stronger the elites, the less people can hope to challenge them. Other things being equal, this makes people more prone to redirect their hatred to weaker groups.

To this extent, politics is the activity of the collectively defensive, whereas economic enterprise is the activity of the hopeful individual. No wonder free-enterprise capitalism produces more growth—but, alas, not much more environmental protection—than state socialism or state capitalism. When people try to improve their lives or make those lives happier, they rarely do it through political means. The odd vote for museum or park bonds is a truly rare and minor exception. This could be considered part of the free-rider problem, but there is rarely any difficulty getting people to unite against perceived enemies; it is much more difficult to get them to unite in favor of something.

MORAL SUASION BEGINS TO ENTER THE PICTURE

Education and moral suasion are, in the long run, far more effective and cheap than regulation. A major education campaign looks expensive compared to passing a law. In fact, it is the law that is expensive. Forcing an unpopular measure on a sullen community is a terribly costly and highly unsatisfactory enterprise. An example was the Endangered Species Act. The U.S. government, without much explanation, began suddenly to protect a vast number of small and insignificant plants, insects, and the like. The methods used were often heavy-handed and ineffective. In hindsight, it would have been better to explain the need to save species and let people figure out how to balance that with their private interests. It would also have been better to spend money on reserves, tax breaks for conservers, and even outright payment to affected landowners, rather than on some of the complex and ultimately unworkable plans that were in fact developed. An "endangered species rent," fair compensation paid by the country to local landowners, would probably be cheaper than the current costs of planning and enforcement.

Internationally, as we try to save for the whole world the species endemic to desperately poor countries like Madagascar and Indonesia, there really is no other solution that will work at all. It happens that the countries with the highest biodiversity are among the poorest on earth. It is not fair to make them pay all the costs of the worldwide benefits we want. They cannot possibly pay them in any case. If this is true internationally, something similar must be true within nations, though considerations of freely undertaken risk and of ability to pay have to be taken into account.

The most important principle is, however, that *the duty of government is to protect its citizens, not to produce material commodities.* Government has the responsibility of protecting livelihoods, but not of creating them. Those two goals are inevitably in conflict, and bitter experience throughout the 20th century has shown that a government involved in production will sacrifice its people to the production process. *Governments must act to prevent risk, to protect healthy environments, and to protect genetic information—endangered species and varieties—just as they act to protect individuals.* This is the basic rule for the future. Economic growth is good, even necessary, but only if it actually helps people. "Growth" that merely damages welfare is perverse. The economy—and the government—exist to serve people, not the reverse. We have to recall Kant's principle, which I believe to be basic to all political matters: People are always an end, never a means. We should now add that a healthy environment, and preservation of species, is necessary to human welfare, and therefore basic—more basic than any imagined economic advantage from destroying same.

Even in the United States, where accountability is legally guaranteed, it is notoriously difficult for ordinary people to exert leverage on big business or big government, no matter how worthy the cause. It follows that we should work toward a system guaranteeing both limited government and limited power of monopolistic firms, and counteracting the disinformation spread by some of the latter (cf. Ehrlich and Ehrlich 1996; Stauber and Rampton 1995).

The advantages of grassroots planning, education, and moral suasion are not solely or even primarily economic or political. (See Putnam et al. 1992 on democracy in Italy.) They lie principally in their tendency to create personal responsibility and social solidarity (as the Founding Fathers emphasized). Encouraging local cooperation is not a totally effective way of fighting this, but it is probably the only way that works.

FURTHER ISSUES

Proposals for a cure obviously have to go somewhat beyond my minimal formula of economic reform, accountability, and recourse. Obviously, an extremely vigilant public, deeply concerned with the issue, is an absolute necessity. Public apathy is *the* enemy of the environment, in the United States and elsewhere. But, worse, in countries like Guatemala and Brazil, environmental activism can be fatal. We must not only cure apathy, but also work for human rights.

Yet the problems are too wide and complex for "political" solutions in the narrow sense. The issues are not only inextricably intertwined; they also relate to a rather wide range of human thought and behavior, from education to international treaties. Politics-as-usual, the business of jockeying for power in the anthills of public life, is not even close to adequate.

Progress toward better environmental management can come only within a free society that respects the individual rights of all citizens. This is another "process goal"; the more rights, the better the environment. We cannot, unfortunately, expect perfection, but any improvement in human rights should normally be expected to help the environment (other things being equal). Freedom, with respect for rights, is the only route to real progress toward ecotopia (Johnston 1997). It leaves us with some substantial questions of "rights" and of "respect," and it may have to be qualified with further learning, but so far it seems the best we can do. This is a fairly widely held view (cf., for example, Ehrlich and Ehrlich 1996; also, "The Earth Charter," *Earth Ethics* 8:2 & 3:1–3, 1997). However, it goes against the opinions of some segments of the environmentalist movement—the ones who favor strong centralized government as the only hope, and the ones who favor a powerful, uniform environmental morality.

There is much literature that unites the question of individual and local rights with that of ecological protection. Some writers start from ecology and move to rights (Ehrlich and Ehrlich 1996; Sachs 1997). Others start from trying to protect local rights, especially of indigenous minorities, and deal with ecological issues as one type of issue involved (Handwerker 1997; Johnston 1998; Maybury-Lewis 1998; Newcomb 1994). These authors have started from various positions, but they come to approximate each other closely. The general conclusion is that ecological mismanagement and environmental devastation clearly damage people, both as individuals and as members of groups (however defined), and these people have rights as human beings to resist. Even Thomas Hobbes (1651), not a great exponent of freedoms or of democracy, started with the English common law position that a person has a natural and inalienable right to defend himself or herself against attack. If there is one human right, this is it. If resistance to direct threat is a basic right, then we all have a right—and, in fact, a collective duty—to resist destruction of our life support system.

A rights-based environmentalism would necessarily include some further rights. To defend themselves effectively in a world of vast productive interests—big firms and big governments—people need the classic democratic rights of free speech, freedom of assembly, and freedom of conscience.

That being said, a problem immediately arises. Advocates of indigenous group rights are especially prone to sweep it under the rug, but it keeps breaking through. The problem can be stated as a question: What if the government has a genuinely good plan—for reforestation, or a wildlife refuge—and the locals are against it for bad reasons? Perhaps they want to cut the forest down and run cattle, perhaps an international mining company has paid them to be intransigent (this does happen), or perhaps they want to hunt local game in disregard to its future extinction.

It is more or less universally agreed that governments have the duty of balancing one person's rights against another's. So do individuals and firms, for that matter—to the extent of their legal ability to do it. As we all know, one person's rights stop where the next person's start. In all modern societies, those rights are equal, according to legal fiction. In ecotopia, they would be so in reality. No one—not ranchers, not indigenous groups, not environmental activists—has the right to damage another without reason or recourse or compensation. I have a certain sympathy with anthropologists who want to "protect" the rights of a small tribe to raise cattle or take fish, but if their actions devastate an environment that is essential to other people's livelihood, they simply have no moral claim. Many anthropologists champion "indigenous" groups. This privileges a highly arbitrary and specious concept of "indigenousness." Many "indigenous" groups are as recent, or almost as recent, immigrants to an area as the "non-indigenous" groups are. The rights of people on the ground *must* be respected, and indigenous groups deserve full respect and support, but they have no more right to destroy the environment than less "indigenous" people do.

Certain Asian dictators have lately been excessively zealous of "group" rights. In practice, this turns out to mean denial of individual rights. The "group" in question always has an amazing resemblance to the dictator and his clique—presumably the rest of "the people" are considered irrelevant, or else an extension of the dictator's person. This argument would not deserve consideration, were it not taken seriously in some quarters. It has been especially visible through the claims of some individuals disingenuously misrepresenting "Confucianism" or "Asian tradition" to the west. Suffice it to say that both mainstream Confucianism and most local Asian philosophical and political traditions involve respect for the individual, and rejection of the sort of tyranny that now tries to seek justification. Mencius, the greatest philosopher in the Confucian tradition, said forthrightly that the people have not only a right but even a duty to throw out a cruel or irresponsible ruler. Would that certain Asian nations took Mencius more seriously today (Leys 1985; Mencius 1970).

Protection sometimes involves a limited sort of production—most obviously when the government reforests a slope to stop erosion and flooding. This introduces the threat of abuse of force, and so it should be done only with the agreement of, and with compensation to, the local people. Today, enlightened governments hire the locals to do the work, thus ensuring that the locals have a stake in the forest and will tend to protect it.

Basic research is a very different activity from afforestation, but it has the same justification: it is a necessity for survival, at least in the modern world, and not a "production" issue. Applied research can be left to corporations. In either case, funders should leave the scientists alone to do what they think best, once a general policy has been set. Peer review works—often badly, but at least it works; pork-barrel support and national fine-tuning of science do not. Similar things can be urged of education. Education is necessary for survival in the modern world, and governments cannot dodge the responsibility for it; the American right-wing attack on public education is absolutely suicidal from a national viewpoint. Most right-wingers now appear to believe that education should be left to private and, especially, religious schools—most of which are small fundamentalist organizations that refuse to teach evolution, let alone other facts of life and biology. But education should be left as much as possible to local teachers and to independent experts. Both conservative and liberal domination seems problematic; high authorities, not dealing with children and classrooms, can make mistakes for even the best reasons. The current boom in standardized testing seems a case in point.

Fourth, and probably most significant, local people manage resources very well indeed, especially when they have been in the area a long time and depend on its resources. They do best when given some support or help. As long as comanagement means expert advice on managing the resource and setting up infrastructures to use it, and as long as the people on the ground have control over the process, it is a good thing; if it turns into subsidy or domination, it has not worked well.

Fifth, there are simply some places where environmentalists will have to grit their teeth and bear it. Local people must have the right to make their own mistakes, and wreck local resources here and there if it does not excessively endanger the wider system. There is an obvious issue of consistency here—if personal freedom doesn't mean personal freedom, what on earth does it mean? There is also a practical reason. People make mistakes. If a small community (or family, or business firm) makes a mistake, neighboring communities (families, firms) learn from it and stop doing it. But if they are prevented from doing that, and the wider society takes the initiative, the wider society may make the mistake. If it is made at a high level, where people are buffered from paying the costs of their bad decisions, then it not only is not repaired—it is usually widely imitated.

Some of the previously mentioned principles come into conflict in the real world. One has to determine, case by case, whether a reforestation project is a general benefit or a subsidy to a particular person. If there is doubt, the precautionary principle applies: Do nothing without proof that it's good; do not wait for proof that it's bad, because that proof can only come through seeing the bad happen. One has to determine, case by case, when comanagement starts to slide off into government control of local people's production. This is a well-recognized and chronic problem. One has to determine, case by case, when an indigenous group's behavior is infringing on another group's rights. But none of these problems is unique to environmental management. They are what civil law is all about. The reader may by now understand why there are so many lawyers.

A rights-based environmental polity is not necessarily adequate. There is, among other things, the problem of translating general human rights into specific environmental rights, and then getting them institutionalized in some sort of adequate way (this matter is discussed at length by Christopher Miller in his book *Environmental Rights: Critical Perspectives* 1999).

In any case, as the proverb says, "the proof of the pudding is in the eating." A mix of reserves, indigenously owned and managed areas, and secure logging concessions that are best exploited by selective logging is saving the Peruvian Amazon (Oliveiros et al. 2007). Areas without such secure tenure and established rights are deforested; areas with them are in good shape. The same is widely known to be broadly true in developed countries, with some sad exceptions, but it is heartening to see the mix working well in a Third World country that until recently was a sad case.

SOLIDARITY

Possibly the most depressing comment on humanity is the well-known fact that solidarity is most easily created by opposition to an enemy. The enemy has to be human, or at least a threatening creature with an agency of its own. Bears and wolves will do, and so will evil spirits and personified natural forces like Old Man Winter. Microbes are more difficult to fight. Politicians who try to harness oppositional solidarity for other purposes, as in the "war on poverty," "war on cancer," "war on drugs," never succeed well. Many societies have made war on nature, most explicitly in Communist China's "struggle against nature," but this too has failed, except insofar as people see nature as a set of genuinely hostile wild animals or evil spirits. Uniting in support of abstractions like Liberty and Justice usually breaks down into factional politics. One hates one's opponents on the floor, not slavery and oppression.

Fortunately, many other solidarities exist in human society. Most obvious and universal is family. This includes not just blood kinship, but marriage, adoption, fostering, godparenthood, and all the other extensions of kinship. Face-to-face networks and communities are almost equally universal and successful. Julian Steward's classic "band, tribe, chiefdom and state" (Steward 1955), and Immanuel Wallerstein's "world-systems" (Wallerstein 1975), emerge as people range more and more widely beyond the face-to-face setting.

Beyond the face-to-face community, however, maintaining solidarity is notoriously difficult. This has led over time to developing a range of mechanisms. The larger and wider the social universe, the more mechanisms it needs. Benedict Anderson (1991) famously described such large and wide social groupings as "imagined communities," and listed a range of devices. The modern states, especially since the rise of nationalism in the late 18th and early 19th centuries, have newspapers, museums, flags, national anthems, government-funded think tanks, and on and on (Anderson 1991).

Counter-state movements must then have their own solidarity (Comaroff 1985). The working people gradually came to see themselves in class terms, and banded together in labor-union solidarity (Thompson 1963, writing at the height of the union movement; one now needs a follow-up to explain how they lost it). The slaves in the American South managed to create a deeply inspiring world of resistance (Genovese 1974).

Above all, especially in dealing with the environment, people use religion as their great creator and maintainer of solidarity. Religion does not exist to "explain" or to invoke "awe and reverence." It exists to hold society together by getting people emotionally involved in their cultural and ethical systems. Thus it is the "collective representation of the community" (Durkheim 1995/1912)—not a mystic union of souls, but a pragmatic, often quite hard-headed, use of ritual and ceremony to get people together, heart and soul. At its best, religion brings the world's greatest art, music, literature, philosophy, and drama together in stunning performances that fill everyone with enthusiasm for the good and the virtuous. "Enthusiasm" is a Greek word for divine possession—the God within—and God, for most humans anyway, is society writ large.

Far from being other-worldly and too good for this realm, religion is often managed quite deliberately for this worldly task. Insofar as religion retains its moral side, it will teach "creation care." It may still be the best hope. It will not, however, be adequate.

Science offers a better view, one committed to the search for truth and the application of facts to real-world problems. It also has a natural bias toward efficiency and economy of means; scientists have to do as much as they can with very limited resources. The problem with science is that it is individualistic and specialized. Even "big science" actually involves quite small teams. There may be 300 physicists on one experimental project, but that is nothing compared to the number of soldiers in even a very small nation's armed forces or the number of communicants in even a very small religious sect. Scientific associations were lauded by the anarchist Pyotr Kropotkin as fine examples of voluntary associations, but they will never involve more than a tiny fraction of the human race, and it is a fraction self-selected to be orderly and responsible. Science creates communities of interest, and communities united to share in its benefits, but does not create solidarity the way religion does.

Inspecting recent history leads to a particularly important realization. People in most of the world really did make major moral and political progress from around 1750 or 1800 onward. This progress was local, and was locally reversed (as in the fascist years in Europe). However, it did happen.

Most basic to it was one particular thing: tolerance for people who were neither perfect nor much like oneself. Today, environmentalists are torn by factionalism and mutual intolerance. Much of this is the result of the 1960s, when confrontation, extremism, and intolerance became the order of the day in politics. The confrontational approach that developed and triumphed during that period has been, at best, counterproductive on balance. The 1970s brought a reaction against excesses of confrontationalism, but the reaction drifted off into impossibly utopian ideals of humanistic psychology and New Age activities, leading often to disillusion and embitterment in the 1990s. Bias returned, and the flare-ups of vicious racism and sexism in the 2008 presidential primaries shocked the country.

We have to do better. We *did* do better—very much better—before that. The triumphs of conservation from Teddy Roosevelt on through the 1950s were based on widespread cooperation among interested parties. Environmentalism is in sorry shape now because of too many disparate goals and too much puritanism and intolerance about getting to them.

Particularly serious is the conflict between aesthetic preservation and wise use. Aesthetic preservation is *not* elitist; everybody wants it when they think a little. But it becomes elitist when it is done without consideration for ordinary people's livelihoods. This we can no longer afford. Similarly, "wise use" does *not* mean unrestricted license to ruin—to take everything and leave none for others—as the "Wise Use Movement" advocated in the 1990s. In the 1940s and 1950s, there was widespread appreciation that the two not only can, but must, go together; Aldo Leopold's *A Sand County Almanac* (1949) remains the classic statement.

The labor movement remains the best example and cautionary tale. It took terrific effort to build solidarity over the decades. The result was a genuinely powerful, successful movement by the 1950s. From the 60s, however, the right wing took advantage of natural divisions, and the rise of confrontational politics and ideological "purity" in the 1960s played directly into their hands. The result is that labor union membership has declined from almost 40 percent of the labor force in the 1950s to less than 10 percent now, with most of those in the public sector.

POLITICS: FINAL WORD

Politics has operated to save the environment in rather special cases:

- When economic need faced a compact community in which people worked together (Anderson 1996; Ostrom 1990).
- When religion backed up perceived need in a less compact community (Anderson 1996). Sacred groves, taboos on killing animals, and similar phenomena are very common worldwide (cf. Callicott 1994).
- When elites or masses wanted their fun, and set up hunting parks, game parks, or national parks. Ducks are as common now as they were 50 years ago, largely because of conservation efforts by duck hunters. Bald eagles and peregrine falcons have recovered their numbers, under heavy care and management. Other, less fortunate species, however much endangered, have not been the result of much public care, and thus have not gotten much attention or funds.

In the modern world, whose economy and ecology is one indissoluble whole, national and international regulation is necessary. This being said, the people who are actual stakeholders in a resource should be the first and last to have a say.

Decisions about a particular fishery should be made by the fishermen and other interested parties that depend on it—accountable, however, to consumers and others who pay high costs if the resource is exhausted. Decisions about a watershed should be made by all interested parties in that watershed (see Aumeruddy 1994 for a typical case study; Pinkerton 1989 for the general case and some examples).

Experience teaches that individuals profiting from an activity never regulate it well enough unless they are in very stable communities. In the vast majority of situations, there is a need to bring all interested parties to the same negotiating table: sellers and buyers, fishermen and fish consumers, polluters and pollution victims, regulators and regulatees. Too much management now is being done either by sweetheart deals between producers and bureaucrats (cf. American forestry and grazing) or adversarial matches between producers and consumers. In general, no planning worthy of the name takes place under such circumstances. Producers, consumers, and regulators simply have to work together continually.

7

Going Past the Land Ethic

Cuzco: I watch children practicing folk dances in the street. A stray dog tries frantically to join the game. They push him away. Finally, one frustrated girl kicks him. He runs off. Immediately, a well-dressed man—obviously hurrying on a serious mission—detaches himself from the crowd, goes to the dog, pets it, rubs the kicked spot, works its paw up and down. He does this for a long time, until the dog is in good shape again.

Then he goes on his way.

Arequipa: Under the Grau Bridge are shacks built in the river bottom. Most of them are sordid, dirty, in poor shape. One—not the best of them—has a beautiful garden, a paradise of roses and herbs. A woman moves through it, caring for it. For whom is she doing this?

THE LAND ETHIC NEEDS UPDATING

Aldo Leopold, in *A Sand County Almanac*, summarized his "land ethic" in a brief line that has become extremely influential. The full passage says, "The 'key-log' which must be moved to release the evolutionary process for an ethic is simply this: quit thinking about decent land-use as solely an economic problem. Examine each question in terms of what is ethically and esthetically right, as well as what is economically expedient. A thing is right when it tends to preserve the integrity, stability, and beauty of the biotic community. It is wrong when it tends otherwise" (Leopold 1949:224–225).

Leopold's call, and the long discussion following it, has never been improved. Most people who have any concern for the environment think beyond economics to ethics. The final part of Leopold's call, however, now causes us some trouble.

When Leopold wrote, the best ecological wisdom held that Nature (capital N) was indeed in balance, stability, and harmony, except insofar as "Man" had damaged it. We now know better (Botkin 1990; K. Anderson 2005). Not only are typhoons, volcanic eruptions, floods, and fires common, they are essential to the maintenance and survival of ecosystems. California's chaparral, the pine forests of Arizona and Florida, and countless other ecosystems depend on fire. The Everglades and many other systems need periodic renewal by hurricanes. (We now renew California's vegetation with controlled burning, but can we order a controlled hurricane?) The Amazon forest cannot maintain itself without floods and windstorms that open the canopy and allow new trees to sprout. Few, if any, ecosystems can maintain their stability without periodic destabilizing factors. Controlled burning, deliberate flooding, and other heroic measures are gaining popularity as management tools. This is risky to people who own houses at the edges of the lands involved. Yet, the alternative of nonburning allows "natural" risks to get steadily worse. Fuel builds up until a vast conflagration is inevitable. The Native Americans managed by controlled burning (K. Anderson 2005). How do we now deal with such questions?

Nor is stability always found. The forests of Canada are still recovering from the last ice age; they have not reached stable "climax" status in 12,000 years. Extinction events may be precipitated by sudden climatic change, or, like the end of the Cretaceous, by meteorite impact. Local devastation by agencies such as volcanic eruption is common.

Not only do humans influence the entire globe; they have been influencing most of it for several thousand years. Africa and Asia, indeed, have histories of human impact stretching through more than a million years. Their ecosystems have developed with *Homo* as one highly influential component. Even the wilderness areas (save perhaps a few ice sheets) are now heavily affected by influences ranging from trail construction to acid rain. Opposing Man to Nature never made much sense; today it makes none at all. Whether environmentalists like it or not, the world is now one big farm. It is the Garden of Eden, demoted, made into fields that are in our charge. The few remaining bits of wilderness—all compromised, if only by airplane noise— are in our charge and on our sufferance. Yet, even now, too much writing on the environment is still grounded in a rather stiff opposition: here are People, and there is the Environment, or Nature.

Daniel Botkin, in his book *Discordant Harmonies* (1990), forever changed environmentalism by taking these facts into account and developing new ideas about conservation and environmental management (see also Borgerhoff Mulder and Coppolillo 2005 for a sensitive, thorough account of conservation based on these perceptions). Obviously, if Nature is not stable, how do we know what to do? Is it wrong in any absolute sense to exterminate species or destroy ecosystems? After all, a comet might hit the earth and do the same damage. If we do want to preserve an ecosystem, how can we do it? All environmental planning is risky and uncertain. Maybe the endangered species are goners anyway and we are losing jobs for nothing.

The new land ethic must therefore be one of managing an unstable world system, not one of keeping our hands off (or almost off) a stable one.

If there had ever been any question that the world is one ecological system, current problems have eliminated all doubt. From whale migration to global warming, we are dealing more and more with problems that do not respect national or even continental boundaries. Even problems that begin as purely local ones come to affect the whole world by disrupting trade or inspiring massive international aid. Drought in the Sahel or typhoons in the Philippines lead to reduction of duck populations in the Mississippi Valley by creating more demand for grain and thus more plowing up of wetlands. China's paving over and polluting of farmland is destroying the Amazon forests, as China is forced to import more and more of its staples from Brazil. The biofuel revolution has led to still more destruction of Brazil's forests.

Indeed, one area's ecological improvement may lead to another's devastation. This happens most commonly in farming; one country's crackdown on heavy agricultural pesticide use often leads to expanding agriculture elsewhere, with even heavier use of pesticides. Strict pollution laws in the developed world have led to exporting industry to developing countries that have no controls. Thus, much local ecological action is no more than downstreaming one's problem.

In short, "think globally, act locally" must change to "think globally and locally, act globally and locally."

I am struck by the extreme backwardness of moral philosophers like John Rawls and Tom Scanlon in this regard. They are still talking about a world where environmental problems do not exist. They are also still talking of a world of rational individual humans, inferentially adult cosmopolitan males. (See Rawls 2005 for some discussion of this issue—complete with rather lame admission of those charges.) Social emergents, nonhuman persons, and sacred spaces are completely alien to their ideologies. They barely nod toward the reality of irrational or choice-restricted humans. The environment is explicitly excluded from their consideration. They do not, and their schemes cannot, deal with the issues and conflicts discussed here.

A final problem with the Leopoldian land ethic is the translation of its economic and esthetic agenda into political will. Leopold trusted in the community, saw ethics as part of community, and wished to expand our concepts of community to include the nonhuman world (Callicott 1993; Leopold 1949). He was raised in a world of small farms and small towns where everyone took care of everyone else. Community in this Leopoldian sense is now a nostalgic dream in most of the world.

We must develop a new land ethic, based on awareness of the inevitability of change, flux, and "discordant harmonies." This will look less like Leopold's ethic and more like the ethic of sustainable resource use advocated by Herman Daly (Daly and Cobb 1994), perhaps combined with a milder form of the "ecology of freedom" advocated by Murray Bookchin (1982). It will, more broadly, look like an "environmentalized" form of Jurgen Habermas' communicative ethics.

In what follows, I will keep citations to a minimum and refer the reader to the "supporting online material" in "Ethics," posted on my Web site at www.krazykioti.com. It contains citations and supportive background material too long and dull to include here.

PINYON JAYS AND CHIMPANZEES

Christine Korsgaard (1996) writes that the need to be human and reflective lies behind morality. Not so. John Marzluff and Russell Balda, in their book *The Pinyon Jay*, tell the story of a flock leader who attacked a goshawk in defense of his flock. The flock escaped; the leader became the goshawk's prey. Jays are not especially reflective creatures, but they are fully aware of the terror of confronting a goshawk, and they are aware of choices and of what it means to make them. This flock leader made his choice (Marzluff and Balda 1992:46–47).

Biology, in fact, has revealed that most (if not all) species of intelligent social animals have simple analogues of human ethical behavior.

Consider Rousseau's "savage." Rousseau was not talking about an imaginary creature. He found his powerful, hairy, wild, yet sociable savage in the travel writings of explorers of West Africa; it was the chimpanzee (Rousseau 1983). Rousseau correctly inferred from his sources that chimpanzees were highly familial and group-oriented creatures, living in rough harmony with their own kind and even taking an interest in (fellow. . .?) humans. He saw their speechless, simple, yet warm sociability as the ground on which could be built the "social contract" that creates human society. (Incidentally, Rousseau never said the chimp was a "noble savage." The phrase was Dryden's, not Rousseau's, and referred to French romanticizing of classical antiquity, not to anything real.)

We now know a great deal more about chimpanzees and other apes than Rousseau did, and it appears his inferences were not totally off the mark.

Frans de Waal describes chimp ethics (de Waal 1996, 2005). De Waal and other investigators have established that the rudiments of morality are widely distributed among social animals, and that chimpanzees and bonobos (as befits our closest relative) come very close to us humans in the complexity and sophistication of their moral judgments.

Chimps appear to have concepts of fairness. They routinely help each other in everything from treating wounds to getting bananas. On a more sinister level, they learn to fight over scarce resources. Chimps can tell "murder" from mere "killing," and their more spectacular murders occur when high-quality food is limited and concentrated.

With chimps and bonobos, it is hard to tell when a behavior is mainly "instinctive" and when it has been worked out because it is reasonable. Apparently, in these primates, genes specify a very high degree of sociability and a general ability to think, plan, and scheme in social contexts. Genes also specify broad classes of responses: aggression, reconciliation, cooperation, and so forth. Many behavior patterns (including much of the sexual response system) seem heavily gene-guided. On the other hand, these apes can choose when to invoke what behavioral strategy; how to fine-tune it to the occasion; and how to combine it with other strategies. More interesting is their ability to use or hide a behavioral strategy when it seems "inappropriate," as a way of manipulating a situation. Most interesting of all is their obvious thoughtfulness about social action. They plan their social moves. They have true politics.

AND PEOPLE: MORAL NATURES

Humans form much more high-level representations than chimps. We can set goals, design overall plans, work out strategies and priorities, and design tactics for actually carrying out the strategies. We can change our tactics in midstream and think up new ones. We can coordinate the behaviors of dozens or even thousands of people. But our moral grounding remains rather like that of the apes. Obviously, we are even more social than they are. Friendship means even more to us, and we have added love in all its complex forms. In short, "(moral) thinking is for (social) doing," in the words of Jonathan Haidt (2007:999; see also Cheney and Seyfarth 2007).

People *care* about others, including animals and plants, not just humans. They have some sort of innate sense of fairness, and they certainly react negatively to others' pain. We still do not know how far this innate sensibility extends, but it is clear that morals are not entirely socially constructed. Justice, fair distribution, a general sense of responsibility, a sense that one should help and not hurt others, a strong sense of freedom as a right, and a sense of conformity to social codes all seem to grow from innate roots. Some sort of group or multilevel selection seems necessary to account for this, but there are models that allow it (Handwerker 1997; Wilson 1998). Morality is constructed on the basis of these primes. Everyday behavior involves constant negotiation (a less radical process than construction) about fairness or justice in particular cases.

Humans also have an inborn intolerance for cheaters and unreliable people generally (de Waal 1996; Hauser 2006; Henrich et al 2004; Petrinovitch 1995). Playground fairness—the worldwide toddlers' rules of fair trade, sharing-not-taking, helping others and not hurting them, and so on—is at least partly hardwired in the human brain. We may even have a mental "moral organ" like the inferred system dedicated to language (Hauser 2006). The old concepts of morality that saw humans as innately selfish, power hungry, violent, or amoral and saw morality as an imposed, rational, "civilizing" force, were simply wrong. Freud, Nietzsche, and the Calvinist codes of my youth have alike lost all their grounding. People are not necessarily "basically good," but they are not basically bad, and certainly not basically selfish.

Humans are naturally more socially aware than even their closest relatives and have an innate sense of fairness, however much it may be manipulated by culture (Henrich et al. 2004; cf. Taylor 2006). Unfortunately, the worldwide ingroup-outgroup separation and tension does seem to have a biological basis (Hinde 2007).

Humans simply do not act as individual maximizers ("optimal foragers") in the sense that mountain lions or even chimpanzees do. There are society mates to consider. Thus, we expect to find, and we do find, that humans almost everywhere manage resources for sustainability over the long term. This has been constructed morally and socially everywhere. It is typically part of religion, thus being enforced by religious teachings, often including both promises of heaven and threats of hellfire or its equivalent.

Similarly, the simplistic explanations of morality and altruism as the result of kin selection or reciprocity are inadequate (Haidt 2007). People sacrifice their lives for strangers and causes and even for abstractions like the flag and the church. People derive their moral codes from huge cultural groups, not from families; if forced to choose, we adopt the morality of our wide peer group in preference to that of our parents (Harris 1998). People react quickly and intuitively, evaluating everything they perceive, usually in some moral way (Haidt 2007); on the other hand, we think constantly about such reactions, though often to justify them rather than to qualify them.

More dramatic are new findings on the role of oxytocin in these matters (see, e.g., Moll and de Oliveira-Souza 2008; Price 2008). People like being generous, helping others, and being what we normally call "good" and "nice." They get internal rewards, in the form of oxytocin release in the brain and other neurochemical reinforcers. This makes the cold-blooded utilitarian calculus difficult and the "rational individual maximizer" impossible.

The more closely an issue is involved with social marking and sociable security, the more emotional it becomes and the more "moral." Our way of doing things is moral; other ways are not. Any rival group's way is the most immoral of all. One recalls, from Richard Brandt's study of the Hopi (Brandt 1954), that "Hopi" is both the name of the people in question and the word for the morally good. "Ka-Hopi" (Brandt's spelling; now *qahopi*) means both "not Hopi" and "immoral." Most people worldwide seem to have the same idea, if less clearly stated.

"Morality is about more than harm and fairness" (Haidt 2007:1001). Haidt lists other dimensions of morality: group membership and loyalty; respect and obedience; "bodily and spiritual purity"; and "living in a sanctified rather than carnal way" (Haidt 2007:1001). As he says, these are more stressed in some cultures than in others. I can add that anthropology has found the first two of those other dimensions to be absolutely universal and sees them as inborn, and the third probably is also (Douglas 1966), but the fourth is definitely a product of modern world religions, found, to my knowledge, only among people strongly devoted thereto. Traditional, small-scale religions were more involving; people did not really have the "carnal" or "secular" options.

Thus our failures of perfect rationality force us to come up with moral institutions—emotional, typically religiously grounded, rules that force us to do what common sense should really require.

In so far as rational choice theory is correct, economics must fail at making people act responsibly. Short-term and "selfish" behavior always wins out because it provides an immediate competitive advantage (Olson 1965; Hardin 1968). Even such things as "tit for tat" games do not survive in an economic world, since the selfish player simply eliminates the unselfish one immediately by economic competition or, in the real world of political power and hate, by assassination. Only an actual, explicit moral code can institutionalize long-term, wide-flung concerns as prior—*or even considerable at all*.

Moral and ethical codes are institutions of a special type. They are absolutely critical for maintaining trust and reducing enforcement costs to a minimum. Note that institutional economics allows us to see morals, and even religious rules, as part of the economic system. This saves us from the unnecessary and unfortunate tendency to separate economics from "spiritual" and "deep" ecology. One can also do this the other way round: see economics as part of the moral system. If you suspect people do work, usually to maximize wealth, you might want to start with economics. If you suspect people work out of love (however defined), you might start with religion. Either way, you come to the same place: an analysis of just how moral institutions function to keep the costs of society to a minimum.

Most people can be trusted to look after their short-term interests. We do not think of buying food for oneself as particularly an ethical issue. Feeding hungry strangers is. Laying in a supply of food for oneself in case of an earthquake is somewhere on the border. It is not quite a social virtue, but it is part of an ethic of being provident and careful. Here, again, morals and rationality come to something like the same thing: ways to calculate long-term interests so that we do not sabotage ourselves by looking solely to the short-term ones. Thus John Rawls (1971) and Amartya Sen (1992) have emphasized fairness and equality of access.

An ethic that privileges long-term and wide interests over short-term, narrow ones must be made explicit in particular cases. An adequate code would also take care of the future: "What possible argument could at the same time require that the present generation have scruples about leaving enough for one another, while shrugging off such concern for future generations?" (John T. Sanders, quoted in Schmidtz 1991:24).

Even if we were all virtuous, we would still need social ethics. Problems come when you try to decide conflicting interests. How do you trade off endangered species protection against jobs, or spending on parks against spending on hospitals? How much care do you owe animals (Sidgwick 1907:414)? Unnecessary cruelty is surely out, but have you the right to kill them for food? Trying to calculate such tradeoffs with a utilitarian calculus is nightmarishly difficult (Elster 1983; Sidgwick 1907:145, 158).

There is the problem of defining who is within the moral matrix: Only one's own group? All humans? All sentient beings? All beings? Cultures? Languages? Some ethicists even want to save analytic abstractions, such as "ecosystems."

When we are not paralyzed by fear that translates itself into scapegoating, we can see a world of beauty, health, and harmony—the three things the Navaho combine in their unitary concept of *hozhoo*. People would want to improve themselves not only in health and environmental security but in access to beauty and caring. This would be a world of personal commitments, mutual aid, and, above all, appreciation of others and their creations.

Anthropological research shows that, cross-culturally, morals are very different from those postulated by grave philosophers. In the real world, or rather the cultural worlds of humanity, we observe the following:

Morals come from practice. They come from people's actual *experience* of what hurts them and their loved ones.

Morals are especially concerned with real life-or-death matters—in these cases, with livelihood. It is more important to protect the resources on which the community's life depends than to protect the merely beautiful. However, in all these codes, the moral teachings *run far beyond the necessary and enjoin protection of the rest too.* On the Northwest coast, especially, unnecessary killing of any animal, no matter how useless, is heinous.

Morals are phrased in terms of the culture's cosmological belief system. They may have the immediate practical function of saving trees, but the ultimate reason why saving trees is good (and cutting them is bad) depends on the cultural teachings about ultimate cause.

Morals have to be taught. Children have to have at least some real experience with the resources in question. They have to know something about proper and improper use. They have to live the experience. Thus it is properly a source of very great concern, among environmental educators, that most children in today's world have very little contact with anything except urban or intensively industrial-farmed environments.

Morals are effective. They work. The commonly held belief that people act according to immediate self-interest, not according to morals, is simply wrong. (Most of the people that make the cynical claim are quite honest and reliable sorts; one wonders if they ever look at themselves. Perhaps they think they are the only moral ones!)

Morals are not perfectly effective, and no one expects them to be. They are ideals to strive for. There have to be sanctions to keep the dissenters in their place. Moral codes vary a lot in their effectiveness, depending on the thoroughness and effectiveness of the teaching.

MORAL SUASION AND INTERACTION

The most basic key to environmental ethics would seem, reasonably, to lie in recognizing that every human is important—that everyone matters, in some sense. Everyone has intrinsic value and human dignity.

Almost all environmentalists, even the most anthropocentric of "wise use" conservationists, would agree that nature also is important. Plants and animals matter. They have intrinsic value and interest. They are not to be destroyed without consideration. Perhaps we may conclude that they do not deserve to survive. I would not mind if malaria and cholera became extinct. But we have to think about this. Major debates have been held over whether the smallpox virus should be exterminated totally; small supplies are now preserved to make vaccine in case of an unexpected, but conceivable, further outbreak.

This sort of giving importance to all organisms is not simply a matter of valuing them for what they can do for us. It is not simply tolerance. It is not love or compassion, either. One can love, care for, and feel compassionate for people without valuing them or regarding them as important; condescending pity for poor children is far too common. We can care in a limited sense for things we do not value, as grad students care for laboratory rats. However, real caring—with all the interest, involvement, concern, and responsibility that entails—requires more. Love is not enough. We have to respect human dignity and the dignity and deservingness of other organisms.

LEVINAS AND INTERACTION ETHICS

As I look out at bushes and trees, listening to a flow of conversation in the background, I am reacting to these at a far deeper and more total level than I let myself realize. They are still shaping me. The main job of laying foundations for "self" and "person" was done more than half a century ago, but—at both conscious and preattentive levels—I am still constantly working on them. Meanwhile, I am constantly working on the building that has developed on those foundations.

Here we turn to the dialogic and religious phenomenology of Emmanuel Levinas (1969, 1985, 1989, 1991, 1994, 1998a, 1998b). Levinas realized that if interaction and evaluation are prior to knowing and being (and they are), then ethics is prior to philosophy, and ethics is interactive, not individual. Not necessarily "ethics" in the textbook sense—rational rules of conduct—but the real ethics that govern our dealings with other people. Our minds are literally created by intense, emotional interaction with those others. We understand helping, harming, and responsibility before we can "think" in any meaningful sense. *Cogito, ergo sum* is the end point in a long process. No wonder Descartes went so wrong (Damasio 1994). He thought, but he had forgotten that far more intense, searing, and deep physical and emotional experiences came before the thought—before the rational, self-conscious, self-reflexive thought that he meant by *cogito*. And the very core and basis of those blazing experiences that formed our lives were the experiences of help, harm, care, recognition, sociability, love, warmth, anger, forsakenness, and, perhaps most formational to "self" hood, the anxiety of abandonment. From the fear of being alone and the warmth of active, warm interest in each other, we construct a world.

For Levinas, the key words are *responsibility* and *interest*. (He meant being interested in others, not money added to capital.) He reminds us of etymology: Responsibility is the ability to respond. Interest is Latin: "being between."

However, this does not imply a particular content to morality. It merely provides a grounding for moral codes. Societies must construct their own moral codes. This is where the debates come in: Kantian versus utilitarian, puritanical versus liberal, and so on.

For deeply religious Jews like Levinas, the priority of emotional, supportive interaction is enough to make us infinitely responsible for the Other.

More skeptical readers, and those from other religious traditions, may need further justification and elaboration.

The infinite, or at least unbounded, importance of others is a rational corollary of our recognition of their importance to us. It is also an emotional response to those who make, shape, determine, and validate our selves and our worlds.

Of course, such spontaneous feelings are not adequate in themselves or we would be as happy as bonobos—whose policy of "make love, not war" produces something close to a hippie utopia (de Waal 1998). Kant grounded ethics in reason (Grenberg 1999). Levinas takes us further. He sees the experience of the other person as a far more shattering thing than Kant did. For Levinas, it is something like a mystical enlightenment. Clearly, he agrees with Kant that people are ends, not means. But they are the ends of a more brilliantly intense program. From this, the reason naturally infers a greater importance to humans. They are not just the ends of action; they are the very essence of our selves, our lives, and our meanings as people.

This ethic solves the major environmental problems. Clearly, humans depend on the environment, and the human life support system cannot be ruined—that would destroy other people. But, more to the point, humans are not the only "others" out there. We owe a great deal to the environment. It, too, defines and creates us. It, too, presents itself to our clearest sight with mystical intensity. We have to save it. That is a paramount end of human action, and it eclipses all other imperatives save the imperative to care about and care for other humans. It also follows that preserving the environment from destruction is necessarily more a concern than preserving minor comforts and luxuries for a few humans. This follows both because the environment has its own importance and value and because a really healthy environment is necessary for human survival. It is not moral to destroy the livelihood of millions to provide a few luxuries for a few hundreds.

Clearly, again, humans near to us are more directly important than humans we never meet; but even the latter are of total concern, because as humans we are totally personally involved in humanity. But we also have to care about rainforests and fish because we are totally personally concerned with their lives. This realization would bring us close to Native American environmental ethics, which recognize trees, animals, mountains, and waters as persons—other-than-human, but persons nonetheless.

Levinas speaks of face-to-face encounters as the basic human act. The "naked face"—the face I see when I drop my barriers and defenses and actually look at and encounter the person in front of me—is the Other in pure, immediate form. Indeed, anyone who looks without defenses or distancing at another person cannot help being absorbed in the emotional tides that sweep through the mind.

This can certainly be applied to the environment. To look at the face of the land is to be emotionally caught up in it.

David Hume made the point that we cannot deduce an "ought" from an "is." However, it seems that biology gives us some "oughts" in the form of our inborn senses of morality. It is impossible to sustain Hume's point in practice. When we see anything, no matter how trivial, we automatically evaluate it as good or bad (Zajonc 1980; cf. Damasio 1994). Some degree of rudimentary morality thus inheres in all perception. Certainly, judging people as good or bad is an obvious given of the human condition. People do it all the time; they cannot help it. We may not be able to deduce "ought" from "is" by rational logic, but we do it anyway, quite automatically.

Following such logic, people obviously have to care for the environment to ensure that others are not hurt. No pollution in the drinking water, no pesticides sprayed on farm workers, no rip-off of indigenous forests. Conversely, however, no appropriation of poor people's land to tie up in tourist parks and no bans on hunting and trapping unless something is done for such people as depend on those activities for survival. This can be worked out on a simple "maximize help, minimize hurt" calculus. It is, in short, a utilitarian ethic. But it is grounded on the deeper ethic of realizing we are all infinitely responsible for each other.

Some might see hyperbole in Levinas' "infinity" and prefer to read "large and unbounded." I agree with Levinas: Infinite is the right word. But even if our interactive experience requires only some large and unbounded share of responsibility, it still is so rich, complex, and foundational that the amount of responsibility must be very large and must grow with further interaction. Usually, interaction means the kind of involvement that leads to love. Sometimes it produces hate, but we are still responsible. Sometimes the responsibility is of a hard kind—we are forced to kill or destroy that with which we interact. Killing in self-defense is only the most dramatic such case. Even eating bread takes the lives of countless wheat seeds and of insects that inevitably were mixed in with them; vegetarianism is no easy out.

An inevitable corollary of his view is that we have an actual moral charge to be as open to experience and interaction as we can be. This does not mean sensation seeking. It means opening ourselves to really seeing the other, face to face (Levinas 1998). If we are to have an environmental ethic, we have to begin by seeing—actually *seeing*, not just looking at—the warblers, walnuts, and waters around us.

We have to cut the social and personal barriers that keep us from such clear sight. This cut requires at least some serious meditation. We have to do the looking—to see those others who are truly and infinitely important, rather than trivial problems and worries or scientists' abstractions and simplified representations. Following Levinas takes serious moral discipline; we have to cut the old simplifications that blindfold us with indifference and see the brilliant lights around us.

One major value of mystic meditation is that it allows us to drop the barriers that keep us from seeing others. Meditation (correctly done) breaks the barriers down, at least for the brief time we are meditating. We see whatever we are contemplating in all its glory, interest, and intensity.

Great art does something similar, but slightly different: It shows us the value of whatever is the focus of the work. Tragedy in particular drives home the intrinsic value, or worth, or importance of the tragic hero—fatal flaw and all. Landscape art shows the landscape in its full glory. The contrast here is with the art seen—for example—in second-rate Hollywood thrillers and war movies. People are killed as if they were disease germs.

Taoism and Zen teach opening oneself to the world through meditative or contemplative clear sight. (I think that one source of the similarity lies in Hasidic teaching stories; they were greatly influenced by Asian traditional religious teaching stories.) The difference lies in two things. First, Levinas does not build from a long meditative discipline that guides the looker to a certain sort of experience. He starts with the newborn baby (if not before) and looks at the whole development of interaction, from raw new experience onward. Second, he does not construct his experiences into a remote, other-worldly code that idealizes isolation and inner experience. He constructs it into a warm, directly human, totally engaged ethic of helping people.

Levinas himself lived a remote, philosophic lifestyle. Yet the revelation of Levinas' ethic is that one can go directly from clear sight of the world to caring and engaged behavior in the world. This may be no more than a sophisticated version of what all normal parents know by experience, at least about their children. But we need the sophisticated version. We need even the unsophisticated version. Desperately.

We cannot expect innate love for life (Wilson's "biophilia," 1984) to be overwhelming, we can certainly expect *some* extension of infinitely important interactions, and thus of moral life, beyond species lines. However, such grounding is necessarily weaker than our concern for each other as humans. Of course, many people love their pets—close neighbors and "family members" of different species—far better than they love human strangers. Many people, moreover, feel more responsible for their pets than for human strangers. The morality of this is complex and involved but clear enough in the concrete case. We may desensitize ourselves to maltreatment of animals, but it really is desensitizing—forcing ourselves to pretend we ignore perceived maltreatment. It is a minor or weaker form of the desensitization that Baumeister (1997) reports for torturers.

However, morality toward other lives must be expected to attenuate to the vanishing point at the margins of familiarity. It is really hard to become deeply concerned over the fate of a bush halfway round the world, or of a gopher hidden in the soil, or of an animal as different from us as a sponge or worm. As de Waal points out, this is as it should be, at least from an evolutionary perspective. What if we were equally concerned about all lives? How would we care for our children and neighbors?

Our dealings with the environment are, clearly, interactions. Ecology, by its very definition, is a study of interactions, connections, flows, exchanges, and relations. Most authors, from Ernst Haeckel (who defined the field) down to modern scientists, have never imagined differently, though a strong and healthy individual-reductionist influence entered also, via Darwinian selection theory. Concepts such as "ecosystem" and "food web," and even "conservation," make no sense outside of a relational grid.

There is no such *thing*—in the literal, physical sense—as "the environment" or "nature"; there is, instead, a vast summation of human dealings with nonhuman things of many types. This is not to say that "environment" and "nature" are empty words. They are useful labels for whole sets of intertwined relationships and complex feedback loops.

When we study "the environment," and our dealings with it, we are really studying a highly complex set of interactions and relationships with other lives. Ecologists talk as if "the environment" and "the ecosystem" were real, bounded, identifiable objects. Postmodernists respond, correctly, that these terms are clearly vague and clearly the product of a long history of contested use but then go on to say or imply that, therefore, there can be no environmental problem—since the environment itself is pure nonsense. This kind of pernicious nominalism can be combated only if we are quite clear about what we mean. What we mean, when we talk about human-environment issues and problems, is that complex set of interactions with actual beings, especially living beings.

UTILITARIANS AND KANTIANS

A question for utilitarians is that of freedom in general and of particular material or instrumental ends. The Soviet Union stood at one extreme on this point; it abnegated freedom for the sake of guaranteeing rights to housing, food, and other material needs. Its lack of freedom led to its failure to supply the material needs, but in any case the citizens eventually found the lack of freedom intolerable, in spite of decades of conditioning to prefer the

Soviet way (Buchowski et al. 1993). The libertarian view is the opposite extreme, putting so high a value on individual freedom of action that society's existence is potentially endangered.

If we reject extreme positions, we are still left with a huge range of possible tradeoffs. We are daily forced to choose between love and work, buying better clothes for our children or donating the money to charity, feeding ourselves or feeding the hungry. How does one compare utilities in these cases? How does one calculate the relative costs and benefits? In the real world, we will never have clear and unmistakable preferences or preference orderings. Also, we will never have clear and unmistakable ethical guidelines to every possible situation.

Unfortunately, the rapid expansion of world population makes freedom increasingly limited. John Locke wrote, "Labour being the unquestionable Property of the Labourer, no Man but he can have a right to what that is once joyned to, at least where there is enough, and as good left in common for others" (Schmidtz 1991:17). In the 17th century, leaving "enough, and as good" appeared simple enough. To Locke, most resources must have seemed inexhaustible: air, water, North Sea herring, and even fertile land. Locke was also thinking, however, of such things as oak trees and deer, which the English government protected for conservation reasons. Such things seemed very minor in his time. Today, water is scarce, the herring are almost gone, the fertile land is going fast, and even air quality must be carefully regulated. A literal interpretation of Locke would make private ownership of natural resources unthinkable in the modern world, since we now cannot really leave "enough and as good" (Schmidtz 1991)—a possibility foreshadowed in Proudhon's claim that "property is theft" and in Marx's socialism.

We can regard as a nonproblem the "rights" of an individual to destroy a whole society's key resources for purely individual reasons. In a moral society, regarding such a thing as a "right" would be unthinkable. A fisherman could not catch more than a fair share of the fish, even to feed his family. Loggers could not even contemplate logging its redwood forest in a one-time-only, forest-destroying way.

We may seek for a Bayesian optimum between maximal social benefits and maximal freedom. Actual calculation of such an optimum seems impossible in practice. The real-world representation of this idea is normally to maintain freedom in all things unless there is a clear and present social need for controls, as with public health and sanitation. Freedom also includes the right to an opportunity and to an overwhelming social benefit. This occurs in the case of public education. It advances the cause of freedom by equalizing opportunity and the ability to make choices and use democratic rights. A person without education is so handicapped in modern industrial societies that she is at a major disadvantage.

Society might resolve (by democratic debate) to maximize sustainable environmental benefits, enforcing them rigorously and zealously. This would require that some bundle of benefits be agreed on. Food, timber, mines for minerals, clean water, beautiful scenery, park trees and flowerbeds, museums of ecology, hunting and fishing, bird watching, and butterfly observing would all be compared. Controls over individual action would be imposed accordingly. Personal freedom would not be considered as a basic good but merely a means to an environmental-utilitarian end. The advantage of this system is its capacity for fast, thoroughgoing response to problems. The disadvantages are the likelihood that a fast response will be a wrong one, that more independence and flexibility would have allowed a slower but wiser response, and—above all—that this would lead to the sort of eco-tyranny widely reported from Third World areas where impoverished local people have been driven off their land to make way for hunting reserves and the like (West et al. 2006).

A softer, more humane ecological state might invoke fairly heavy taxation, and also planning mandates, for the purpose of supplying environmental amenities, such as parks and flower gardens. Governments would supply many museums and schools, publish relevant literature, and make space available for meeting of nature groups. A concern less familiar to modern urbanites would be equalizing access to rural and wild environments. At present, less affluent urban residents are denied access to such amenities. There is evidence that contact with green and healthy environments has beneficial psychological effects. It also motivates people to want to maintain such environments. Thus common fairness should make us give a high priority to providing such access (Louv 2005; Sen 1992, 1999).

Even this more statist option would have to make room for constant debate and for "whistle blowing" on any signs of overuse. This requires maximum freedom of speech and even its encouragement by means such as public letter forums and talk shows. Other reasons for maximizing freedom include the need for diversity in resource use, the benefits of free markets in some amenities, and the need for each person to calculate his or her own most optimal way of using the environment wisely. Nothing but disaster ensues when Big Brother—be he government, religious establishment, giant corporation, or literal elder kin—attempts to decide wisely what is optimal for any given individual.

RELIGION

The enormous importance of traditional religions in maintaining traditional environmental adjustments was stressed in Chapter 2. What to do about this in the modern world is somewhat outside the scope of the present book. (I am working on another book on this matter, in collaboration with Kimberly Hedrick.) But one must inquire, at this point, what we can do to bring together emotion, esthetics, and motivation in the way religion used to do but rarely does today. Religious groups are increasingly involved in conservation and environment, from Thailand (Darlington 1998; Sponsel 2001a, 2001b) to the Earth Ministry group here in Seattle. "Creation care" is increasingly used as a phrase that appeals across sects and faiths.

Secular organizations like the Sierra Club and Audubon Society have drawn on or reinvented the basic concept of uniting those emotional aspects of life in the service of environmental protection. They have had considerable success—far more (in my experience, at least) than the more formal and scientific organizations or the very broad, shallow environmentalist ones. They have generally avoided the excesses of preservationism critiqued in this book. They represent a trial-and-error re-creation of emotional solidarity.

I would love to be able to create the perfect organization, but the task is beyond my scope, so far at least. It requires a far more charismatic and perhaps more devoted leader than I. At present, I can only support all movements, religious or secular, that are moving in the right direction.

8

New Moral Codes

When the great Hasid, Baal Shem Tov, the Master of the Good Name, had a problem, it was his custom to go to a certain part of the forest. There he would light a fire and say a certain prayer, and find wisdom. A generation later, a son of one of his disciples was in the same position. He went to that same place in the forest and lit the fire, but he could not remember the prayer. But he asked for wisdom and it was sufficient. . . . A generation after that, his son had a problem like the others. He also went to the forest, but he could not even light the fire. "Lord of the Universe," he prayed, "I could not remember the prayer and I cannot get the fire started. But I am in the forest. That will have to be sufficient." And it was.

Now, Rabbi Ben Levi sits in his study in Chicago with his head in his hand. "Lord of the Universe," he prays, "look at us now. We have forgotten the prayer. The fire is out. We can't find our way back to the place in the forest. We can only remember that there was a fire, a prayer, a place in the forest. So Lord, now that must be sufficient."

Shmuel the Tailor, Venice, California, ca. 1970, remembering his days in Chicago; Myerhoff 1978:112

DO NO HARM

In this chapter I shall completely abandon objectivity, and simply lay out my personal agenda.

We have to preserve what is left of the truly wild, for genetic refuges if nothing else. However, most reserves and parks being established today are not wild. They have been managed for thousands of years by local groups, and are in good shape today *because* of this management.

This preservation involves loving nature and seeing our role as managers, or gardeners (Janzen 1998), in it. We have learned, painfully, that one-size-fits-all conservation and management is always absolutely disastrous. Nature doesn't do it. Darwin showed, and explained, the incredible diversity of nature, and the necessity of such diversity for a stable, functioning ecosystem. The same is true of human management strategies. We need to draw on the variety of successful strategies that traditional communities have elaborated. We need to work out many more new ones.

This is a truly Biblical charge; we are given stewardship in the garden, and, following St. Paul, we must use our "gifts differing" (Myers 1980) to do that.

One may reasonably start with the old doctors' watchwords: "First, do no harm." A bare minimum of morality is that we should do no harm to any person, or by extension any sentient being, unless—at least for utilitarians—the harm is balanced by an obviously and unquestionably higher amount of good done to others. The founding principle might be "help not harm."

Russell Hardin has identified another problem: "Moral theory has become primarily the province of those who inhabit philosophy departments while political theory has become that of those who inhabit political science departments" (Hardin 1988:11). For this and other reasons, moralists become too abstract and individualist, whereas political thinkers run the risk of forgetting morality altogether.

So moral codes should be grounded in something like Levinas's concept of infinitely important interactions. Yet, few discussions have followed up this perception in defining the bases of an ideal moral code. It is based not on rational calculation, but on love, care, interest in others, forgiveness, generosity, and tolerance.

This entails a morality of responsibility, mutual aid, integrity, trust and trustworthiness, commitment and loyalty. Nonviolence, tolerance, self-control, and above all valuing others define the necessary emotional self-management.

An environmental ethic has to be based on mutual aid, compassion, and considerateness. The code would recognize not only that humans naturally and normally privilege immediate concerns over remote ones, but also that environmental problems always seem remote until desperate and hopeless.

The greatest evil, in this moral system, would be deliberate cruelty—gratuitously hurting people or animals, or even plants. To my knowledge, that is indeed the way evil is defined in all real-world moral systems, except that the powerful are often allowed (or even supposed) to hurt the weak. (Here I parallel Rawls's ethic of fairness, but otherwise I differ strongly with his highly rationalist ethic. Many of my reservations are addressed, and my considerations paralleled, by the Christian theologian Celia Deane-Drummond, 2006, in a very thoughtful analysis of Rawls's limitations and the superiority of Christian love and compassion ethics in asserting justice as virtue and help for all as obligation.)

SOLIDARITY REVISITED

This all implies an immediate morality of solidarity. Full support and empathy for all humans, and for all organisms in so far as we can manage it, is a *sine qua non*. Caring and concern are the basic terms—the source of responsibility. I doubt if any environmentalist has ever disagreed with this point directly, but many talk as if they feel the future world order has no place for capitalists, males, nonrecyclers, etc. More common, and at least as wrong, is a narrow individualism. Too many environmentalists think that individual actions can solve the problem. No, the problem is social and political, and can only be solved by a movement that would unify people in solidarity with a common cause.

This would, among other things, produce new leaders. A world of competing or isolated individuals does not produce leaders, and the lack of leadership in the United States and elsewhere is proof of this. What it does produce is mystical and quietist options—basically, good people fleeing from society—or else the more hopeful but no more useful myth of the rational individual exercising rational choice. Both quietism and rationality have their places, and certainly we would not want to get rid of rationality at a time like this, but they are not the major needs today.

All this brings us back to the Levinas ethic of respect and care. It has some corollaries. First, it shores up the high value on learning and knowledge that we have already discussed. The advancement of knowledge, in persons and in societies, is a vitally important corollary. It is also a reward of valuing diversity.

Levinas's ethic also implies a morality that favors positive-sum games and is absolutely intolerant of negative-sum ones. (Zero-sum games are not exactly good, but they are often inevitable.)

Finally, it means that we should put a high social value on inventiveness and creativity. These maximize diversity and teach appreciation for it.

The guiding moral principle for the future should include, foundationally, the realization that *we are all in this together.* There is no escape any more, and there are no more structural opponents. The world is in one mess, and has to face that. We may not be Christian enough to love all humanity, or even all of our neighbors, but we have to face the fact that failure of solidarity now means planet death. Love continues to be the ideal—the gold standard. If only we could all love one another. But if we can't, we have to pretend. In the world, we have to act responsibly, as if we are taking care of our loved ones.

This life-raft morality is easy to understand and fairly easy to sell. It does not require assuming that people are good or that everything will work out. Only aggression and flagrant irresponsibility really breaks it.

This leads us to the next principle: the old American adage that *my rights stop where yours start*. Drawing the line, so that maximal fairness and respect are achieved, seems to capture a good deal of the essence of the Confucian, Buddhist, and civil moralities above. Currently, freedom too often means the freedom of the rich to do anything they want to the poor.

The effector of all the above is the old Enlightenment ideal: liberty, equality, solidarity (once *fraternity*, but that sounds sexist now.) Liberty means all the U.S. constitutional freedoms, including contract and conscience. Equality means *real* equality before the law, without economics or ethnicity prejudicing the procedure. We do not have anything remotely like equality before the law in the United States, especially in environmental matters. Money talks too loud.

RESPONSIBILITY AND RULES

Any society based on mutual care and mutual responsibility will have to have a few absolute rules, but they can be negotiated locally by communities on the basis of the general principles.

Some issues are easy. An environmental ethic has to start with regulating the environment to allow it to support life at some level of adequacy. We are spared from the problem of determining happiness here; we can set a much lower goal, an environment in which we and our children and grandchildren can survive and perhaps seek happiness. Minimally, most people can agree on the need to preserve life and livelihood.

Consequentialism tells us to act differently in the real world from what we might do in an ideal one. In the best of all possible worlds, we would presumably not graze wild lands, and we would not have imported alien weeds. In this imperfect world, we have to deal with established grazing rights, traditional cultures based on those rights, and introduced grasses that can sometimes be controlled only by grazing. Some pollution, erosion, and depletion will have to occur in the process of making a living for seven billion people. Without a utilitarian calculus—direct and constant recourse to material consequences— environmental ethics are fairly empty and vapid. Other issues, such as justice for animals or fairness in providing beautiful scenery, may be added to the material calculus, but the material calculus is always and necessarily there. The environment is material, so helping and not hurting it is a simple and literal matter.

However, ecological destruction, though frequently forced by economics, is often the product of sheer irresponsibility, or of passing on externalities.

If a firm kills people by releasing insecticide into their drinking water without informing them, the responsible administrators of the firm have committed murder, by all reasonable moral standards; the act is no different from a fatal drive-by shooting. The law has not yet caught up with this obvious truth, but at least the moral concept seems clear: People have rights to life and health, and these override any rights that a business may have to pollute. Some businesses have insisted on being paid to reduce pollution or to refrain from building houses in floodplains or on prime farmland. This seems mere chutzpah carried to extremes. Cap-and-trading pollution rights may be a good idea (I think they are), but only for pollution that is not dangerous to life. The argument that "the economy needs" unregulated toxic waste disposal could be used with equal validity to support Mafia hit men.

In the case of public goods, such as old-growth timber on National Forests, according to every standard of morality ever taught in the world, destruction for all time of a vitally important resource cannot be justified by appeal to a small and short-term economic advantage. Conversely, locking up a usable, sustainably managed resource, without overwhelming reason (usually, saving endangered species), is also immoral, given the enormous need out there in this impoverished world.

The case of private property is more vexed. Clearly, we want to stop destructive behavior. Equally clearly, we want to provide maximum opportunity for private property holders (especially if each holder owns a very small share of a resource) to make a living and to manage the resource as he or she sees fit. This is the situation where a moral code is necessary, because heavy-handed legislation is sure to fail. All ecology teaches us that local property-holders, to maintain sustainability, have to have flexibility and the ability and knowledge to act fast and decisively. (This point has often been made by the libertarian conservationists, such as John Baden or Peter Huber.) They need also to value knowledge of plants and animals, so that they will know enough and care enough to make informed choices.

Highly moralistic dialogue in the late 19th century led to the formation of the United States Forest Service, the Bureau of Land Management, the Bureau of Fish and Wildlife, and other institutions. These worked well for a long time, until growing economic pressures and declining public political morality led to erosion and collapse of regulation, especially after 1980. Many of the right laws are in place, but enforcement of them has weakened. (In some cases, such as the Multiple Use codes for national forests, enforcement was lax and uneven from the start.) Cuts to funding for patrols have reached the point at which huge marijuana plantations flourish in remote parts of National Forests.

In the Canadian Maritime Provinces, the Canadian government was aware of overfishing, but for several decades was too generous to existing interests (domestic and, alas, foreign) to impose meaningful limits. The fish are now gone, and the Maritimes have become a giant welfare economy (see, e.g., MacGarvin 2002). Hundreds of thousands of people, intensely proud of their centuries-old heritage of self-reliance, independence, and hard work in harsh environments, have become welfare clients or waiters in tourist hotels. Others have left for the new resource boom, the oil sands of Alberta. These sands too will be exhausted. By any reasonable moral standard, the Canadian government did not act well. Yet it is making the same mistake in western Canada's oil sands, fisheries, and forests.

Utilitarian arguments for the rainforests (medicinal herbs, forest fruits, and nuts) are effective, but what are we to make of the pebble plains of the San Bernardino Mountains of California? These unique plains (covering only a few hundred acres; this paragraph is based on my lifelong experience with them) support a number of endemic species, all tiny, dull-colored, obscure, and of no conceivable economic use to anyone for any purpose. Even biologists have shown very lukewarm interest (I am sorry to say) in these unique plains. Much of the pebble-plain habitat has been developed for cabins, roads, and off-road vehicle recreation. Why not sacrifice the lot, losing a few plants and bugs but making a good deal of money and adding to recreational pleasure? Fortunately for the pebble plains, there is plenty of other land in the mountains, and the Forest Service has protected such pebble country as is under their ownership. The moral issue, however, remains.

Once a species is gone, the unique, distinctive, and irreplaceable information contained in its genotype is utterly lost to the world. Its realized and potential benefits, beauty, unique qualities, and interest and lessons to the world are gone forever. Nothing like it can ever be created again. It would seem that species deserve even more concern than paintings or archaeological treasures, which can theoretically be re-created or imitated.

If this line of thought is accepted, it follows that maintenance of biodiversity and preservation of species and habitats is qualitatively different from, and more important than, any other conservation or environment issue. Systems of reserves are the only hope, and though they have high start-up costs they are not necessarily costly thereafter; they often (but far from always) pay for themselves, through tourism.

Recall, however, that major issues of environmental justice occur if people have to be displaced to make way for reserves. If the people are poor and have been sustainably using the resources, the only course is to leave them to manage, hopefully with help and advice but not domineering from outside. Biodiversity protection does not always pay off for the landowner in the long run. It is often pure cost. Thus, landowners are under extreme pressure to prevent discovery of endangered species. Biologists have been run off Texas ranches at gunpoint. Wild patches of California landscape have been bulldozed for no reason other than the fear that endangered species might be found among them. Clearly, this is counterproductive. It is immoral, by any standards, to force landowners to pay an enormous cost for an uncertain and diffuse benefit.

CITIZENS

Political scientists have recently made much of Alexis de Tocqueville's distinction between subjects and citizens. Subjects are the passive victims of oppressive or hierarchic societies; they do not do much for themselves. Citizens are free agents in more democratic societies; they are fond of starting mutual-aid groups, speaking their minds, and otherwise acting as if they had a say in matters.

Commitment is a question rather neglected in the environmental ethics literature, but is of enormous importance for environmental protection. For most people, worldwide, it is natural to defend one's family, friends, and community in a way that it is not natural to defend a wilderness or an endangered species. If it is notoriously difficult to get men to commit to marriage, how will we get citizens to commit themselves to protecting the environment? They must not only pay lip service to the ideal; but also be prepared to sacrifice a great deal for it, as have the many game wardens slain by poachers. It seems hard to deny that a large part of America's current social malaise is due to an enormous decline in the levels of responsibility and commitment; perhaps the neglect of these concepts by philosophers is somehow related.

Rational self-interest, morals, and laws must support each other. No one of the three is adequate to regulate environmental management (or anything else for that matter). One goal should be to design social institutions that bring all three as close as possible, minimizing contradictions that force people to act against their perceived interest or to conform to immoral laws or illegal morals.

REVERENCE FOR LIFE?

Religion and other methods of maintaining solidarity through ritual and emotional appeal deserve attention as counterbalances to politics in the narrow sense. Religion can be either a unifying force or a dangerously divisive force, and it remains for the people on the ground to work out just how this is to be done. Currently, religiosity has been rather divisive in the environmental movement; radical attacks on Christianity and the promotion of badly misunderstood native spirituality have not won friends, and have made credible an anti-environmental backlash among some Christians. It is time for an ecumenical movement.

There have been many broad and general solutions required, ranging from ones that do very little to change the existing system to ones that totally throw it out and call for creating a new human being from the ground up, complete with a new ecological or feminist-environmentalist religion (Merchant 1992). Unfortunately, we cannot change humans. And in this late and secularized day, it is not really credible that we can start a new religion. Nor are ecological scientists and environmentalists the most obvious ones to do that, if it were possible.

The animal rights activists have proposed a general reverence for life as a grounding principle. To my knowledge, the only group that has tried seriously to live by this principle is the Jain religious community of India. They do not kill insects and do not eat fertile seeds or eggs, or whole plants, let alone animal products. Priests of this community wear masks so as not to inhale insects, and carry brooms to sweep insects out of their paths as they walk. This could be a real start for an ethical system, but it would lead rapidly to hard choices.

Animal (and plant) rights come into direct collision with ecological and environmental morality. Here the question is one of the degree of good that would follow from some immediate harm. This question has arisen in actual cases (see also Callicott 1995), especially those involving introduced animals that were devastating ecosystems. Ecologists want the animals removed, which often can be done only through killing the animals. Animal rights activists resist this method (Milton 2002).

Some activists even think that keeping domestic animals is tantamount to slavery. One may wonder what would happen to the world's livestock under a fully animal-rightist regime. Would they simply be turned loose to multiply and destroy the many areas to which they have been introduced? Would they be sterilized and allowed to graze until they die?

In all these cases, judicious elimination of some animals is necessary to save whole ecosystems or ways of life. A world of rigidly applied animal-rights laws would be a world of catastrophic die-offs of human and animal populations. Once again, rigid deontological morality is destructive.

Animals (and, still more clearly, plants) do not have *rights* in the sense that voting members of a polity do. Animals cannot exercise free speech, vote, or sue at law. They do, however, have rights in the sense that babies and severely mentally handicapped humans do (Midgeley 1995): They are entitled to consideration because they are part of our community, because they are living, and because some, at least, of them can feel and think. Kant and virtually all writers on the subject have agreed that acting cruelly or callously toward animals dehumanizes and debases the actor, who becomes—at the very least—a more cruel, callous person. Any doubts I may have had on this subject were removed by my encounters with certain psychologists who experimented on monkeys and apes. Highly educated, civilized persons, who were members of academic communities, fell into practices that were really quite appalling (and not just to animal lovers). It is to the enormous credit of the animal rights movement that such experiments are now emphatically a thing of the past at all major research centers.

Problems of overcontrol and denial of rights seem inevitably entailed by the strict and extreme moral and religious codes advocated by deep ecologists, spiritual ecologists, and animal-rights activists. It is hard—in fact, impossible—to imagine these without a high priest or panel of religious leaders to determine the right. Anyone who doubts this need only read the jeremiads of the leaders of these movements, or watch the behavior of animal rights activists. We could be in for all the horrors of centralized religion: inquisitions, burning of heretics, religious wars, and, at last, property concentrating in the hands of the religious elite and thoroughly corrupting it. The worst thing that could possibly happen to the environment would be the world victory of such environmentalists.

PROCESS GOALS

Beyond the simpler questions come process goals. These are things we can never totally achieve, but can greatly benefit by partially achieving them. Health is the classic example; we can never be *perfectly* healthy, but the closer we get to it, the better. Knowledge is another example. It's absurd to want to know everything, but (*pace* Alexander Pope's much-quoted aphorism, "A little learning is a dang'rous thing; Drink deep or taste not of the Pierian spring" [from *An Essay on Criticism*]) knowing more is better than knowing less. Similarly, loving more is usually better than loving less. This is especially true with regard to the love of the nonhuman ecological world. Loving individual humans can, notoriously, be carried to excess, but we are not likely to love nature too much. (Hating people because they disrupt it is another matter; loving nature certainly does not and should not imply that.)

Democracy, justice, and environmental concern are process goals: more is better; perfection is impossible. One of the most absurd absurdities of postmodernism was its dismissal of the search for truth, justice, and democracy on the grounds that we could never achieve total success, as well as to stop breathing because we can never have perfectly clean air.

In environmental politics, bigness (of *any* institution), control, and any and all uses of the environment have to be kept within limits. Environmental regulation, even the best-intentioned sorts, needs to be kept in moderation too.

One obvious process goal is the search for truth. Scientific education is a good, and so is scientific investigation by any means possible. Recently, science has been criticized widely. Max Weber might say that, after science and the Enlightenment *disenchanted* us from religion (Weber 1946), postmodernism has disenchanted us from science. Some of the criticisms are just silly, such as the idea that science is a white male supremacy game that has nothing to do with truth. On the other hand, modern scientific *practice* (as opposed to science in general) has its problems. It has recently tended to overvalue lab work as compared with field work; linear as opposed to other methods of generating hypotheses; and increasingly narrow and specific agendas rather than basic theory.

Yet, science continues to provide the best way of getting nearer to truth. A nearer and nearer approach to truth, and a more and more specific focus on it, is all we can expect. Absolute truth belongs to God alone, and should be left there.

FIGHTING HATE AND INTOLERANCE

The worst problem of our time—the 20th century and after—has been social hate. The only cure is the classic sense that "we're all in this together." Social hate grows from social rejection, especially indifference to the poor and unfortunate, who are now forgotten and ignored. Conformity and passivity also result from social rejection and intolerance; people fail to take responsibility or to learn from others. The environment is the worst sufferer from hatred and rejection, precisely because only unified and concerted action can save it. Both utilitarian and Kantian ethics are too limited. Neither of these foregrounds solidarity. Both, in fact, foreground the rational individual. Here I may borrow a bit from the communitarians (for a change) and work a strong sense of local grounding into our new ethic.

One needs to be open-minded about ultimate utility and about immediate rules, but open-mindedness has its limits. Tolerance also is a virtue, but not when it involves tolerating child molesting, or genocide, or any other gratuitous harm. Ultimately, the limits of open-mindedness, tolerance, freedom, and other social goods have to be defined with reference to the all-in-this-together principle, and specifically its higher-level goals: help not hurt, best for all, my rights stop where yours start.

A limited view precludes the wide, long-term vision necessary to ecological management. It also exacerbates hate. If people are busy defending themselves as isolated individuals, they will view the world in terms of opponents. We can no longer afford this division. The best predictor of sound ecological management, around the world, is community stability. The communities that manage well are those that can say, "We will probably be here, together, 50 years from now." The second most important factor is tolerance.

In fact, tolerance, as too often phrased in modern rhetoric, is an incoherent concept. It includes tolerance for intolerance—which is obviously a contradiction in terms. It too often includes tolerance for evil, or, if not, it fails to explain why not. The world community tolerated genocide in Serbia for years, and now tolerates genocide in Sudan and religious war and repression in a dozen countries. Hatred should be intolerable, and violence done purely for hatred should be immediately stopped by the united force of all nations of the world.

Similarly, tolerance for harmless sexual behavior between consenting adults is a very different thing from tolerance for molesting children, taking child brides, and directly harming helpless persons.

Tolerance should therefore be saved for the moral position that is the opposite of bias, prejudice, group hate, and bigotry.

The other sort of tolerance—putting up with actual harm, from child molesting to genocide—deserves a specific moral name. *Copping out* and *putting up with evil* are terms commonly in use. *Denial* is the appropriate word in many cases. Neville Chamberlain's infamous accommodation of Hitler was called *appeasement*, and seems to have been a mix of excessive tolerance and outright denial.

Yet a third form of tolerance also deserves a new name, but this form of tolerance is arguably better than the baseline one. Best of all virtues, it is one that lacks a name. We may refer to it as *valuing diversity*, or as *positive* tolerance. It is the idea that one should evaluate other views, ideas, and cultures according to what they have of good to offer, rather than merely tolerating them—let alone rejecting them because they are different, or because their *bads* are the only things evaluated. *Tolerance, multiculturalism*, and *accepting diversity* mean no more than "All right, I'll put up with the SOBs if I have to." This has its points. We need it. But we need at least as much to look at what good every person and every culture has to offer. The world needs every good idea it can possibly get, and even all of them together may not be enough to save us from catastrophe. Rejecting or ignoring any idea or creation, let alone any living person, is a cruelty we cannot afford. Positive tolerance therefore is the prime virtue of the age. To my knowledge, Johann Herder was the first to argue at length for the idea that we should not only value others' traditions as much as our own but also respectfully learn from them (Herder 2002). Modern arguments for this position seem to trace back to him, so far as I can determine. Herder's turgid prose is one reason this idea has not been more widely attended.

One does not cooperate with those one despises. One does not see any injustice in robbing or killing those one considers fundamentally unworthy. In a civil society, there will be enough differences (in class, ethnic group, religion, lifestyle, or whatever) to make intolerance, or even lukewarm toleration, suicidal. People in an intolerant society will hang together in the event of war (outside threat unifies almost everyone), and may provide minimal justice and crime protection, but they will not be able to make the commitment necessary to sound ecological management. Management requires a social contract of a special kind: People have to commit themselves to sacrificing self-interest (at least sometimes) to the interests of the wider community in the uncertain future. People don't do that for people they don't respect.

Ultimately, disrespect moves through dehumanization to outright cruelty, and cruelty—deliberate harm as an end in itself—is the most absolutely intolerable of all acts in a functioning society. (Acts, unlike people, can and sometimes must be beyond the pale of even minimal toleration.)

Conversely, mutual respect and mutual aid drive egalitarian, grassroots, bottom-up organization and accomplishment against the extreme over-hierarchization of modern life. Second, it builds (indeed it defines) community—it brings people together in the way that matters most. Third, it is an optimal way to organize most environmental and ecological work.

We can distinguish three different moral failings, or sins, here: general irresponsibility and bloody-mindedness toward others in general; sheer indifference toward particular groups, and active hatred of particular groups. The first expresses itself environmentally in littering, destructive off-road vehicle riding, and the like. The second lies behind the indifference toward the poor that allows them to fall such into poverty that they destroy the environment out of desperation. The third is seen in civil wars, dispossession of indigenous groups, and the like. Obviously, today, morality has to set its face especially against group hatred.

In the economy and in all other social institutions, the great moral need is for accountability and recourse. I need only repeat (Anderson 1996) that the interested and concerned citizen *absolutely must* be in a position to challenge, quickly and legally, any and every decision or action relating to the environment. *The very first rule of any ecotopia is absolute guarantee of the rights to sue, blow the whistle, and debate decisions in public hearings.*

LAYERS OF MANAGEMENT

One could perhaps classify environmental ethics in terms of layers. There is a first layer that has wide support: We must prevent environmental damage that is directly life-threatening. Massive pollution by deadly chemicals, and massive destruction of a vital food-production system, must be prevented. It would seem hard to challenge this view, but many in the U.S. Congress do not accept it, and neither do most Third World governments.

A second layer would involve situations in which a very simple utilitarian calculus would be enough: less directly life-threatening pollution and resource degradation. Currently, various claims—from jobs to freedom—are used to counter or defang the argument from responsibility and prudence, so even this elementary moral ground is not well established.

A third layer would involve cases in which serious questions of cost-benefit accounting would be involved, but the utilitarian calculus would still be enough: mild pollution, endangered species of possible but unproven worth.

A fourth layer would be separated from the above by a sharp break. Questions about individual animals, about levels of responsibility to species in general, about really hard tradeoffs (poor people versus parks), and about rights of nonhumans must be discussed much more than they have been. We do not have clear guidelines on the rights of nonhumans. Why should we save, for instance, an endangered brine shrimp? There is really no possibility that this brine shrimp will ever be good for much. Nor is it intelligent, cute, impressive, or otherwise charismatic.

A fifth layer involves questions of aesthetics.

Humans being creatures of both passion and reason, all five layers are always present in us, all important, all blended together. But, for political arguments, we will probably find that the first layer is more persuasive to a wider audience, and the others progressively less compelling. Yet, we cannot ignore even the fifth layer, even when debates seem to turn on narrowly utilitarian ends. The old socialist cry for "bread and roses" captures the most basic truth about the human spirit, and we can never neglect either the bread or the roses in understanding human responses to the environment.

ENJOYING

We want to save beautiful views and lovely flowers. We want to guarantee our children the chance to enjoy a wilderness experience. The problem is that most Americans, materialist and money-oriented as ever in spite of Thoreau and Leopold, object to saving anything for mere beauty if it could be used for even the slightest amount of money. This appears to be as true of liberals as of conservatives. This has often led to sacrificing lands that are not only beautiful but also economically very valuable for watershed protection, buffering, absorbing pollutants, and so on.

We need, then, an injunction to have as much pleasure as possible (within the previously mentioned constraints). This may seem odd for a number of reasons. First, most people believe they are pursuing pleasure already. They are, however, usually pursuing (or merely falling into) conformity, which is easily confused with enjoyment in so social a species as ours. Second, it seems strange to modern Westerners to talk of pleasure-seeking as moral. Yet, the Greeks (including Aristotle in his ethical writings) discussed this issue at length (Hadot 2002). So did the Chinese. One of the biggest problems in trying to save the environment has been public lack of willingness to act out of sheer love and delight in nature.

We all love nature and think it is beautiful. Unfortunately, in my opinion, many of us suffer from a Calvinist heritage that inhibits acting on such grounds. Centuries of religious and quasi-religious Puritanism have caused morality to be identified with misery, or at best with hard work for money, in the American mind. This is ecologically disastrous. The deepest need today is for people to enjoy their environments and thus be moved to save what is enjoyable therein. Puritanism has given us a limited and unhealthy diet, a preference for parking lots over beautiful landscapes, a fondness for complaining about evils rather than fostering good things, and above all a pervasive feeling of guilt about enjoying birds, flowers, and nature in general. Consequently, many people become worried and depressed, and will enjoy themselves if, and only if, socially pressed to it. When outdoors, they feel they have to mess up the environment for some practical reason. Letting ourselves enjoy nature requires some personal effort.

The puritanical urge to impose every detail of one's lifestyle on others is abundantly present in many radical ecologists, who seem to think that the world will be saved only when everyone wears, eats, feels, and says the same thing. The reverse is true. Variety is necessary for maximally efficient use of the ecosystem. If we all ate soybeans, the world would look like central Indiana—not an ecologist's paradise. If we all liked to hike in the Sierra Nevada, it would look like a cow pasture; it looks overused enough already. If we all had to wear the current ecologically correct clothes, the places that produce the fabric would be farmed to death. (Compare the well-meant over-regulation of hate speech. Hate speech can be dealt with only by answering it. Suppressing it drives it underground and prevents open rebuttal, and thus greatly favors the hate message. It is also inevitably taken by the hatemongers as proof that they have something really important and scary to say.)

If beauty, enjoyment, and love are their own reward, and if hate is unpleasant, a fighter against evil would expect more results from flowers and songs than from sour Puritanism. If diversity is the soul of ecology, then the rigid, dour, conformist codes of radical ecologists are soulless.

There was once a concept of the human spirit: the ability to feel deep and intense emotions, to love, to be socially responsible, and to grow from infancy onward rather than shrinking intellectually from the open, active mind of the child. This personal ideal of becoming deeper and more self-aware seems to have died with the rise of television. Nature was once thought an ideal place for such self-cultivation (on this and related issues, see Louv 2005). If we cannot recover this aspect, we can perhaps try to construct a world where interactions of people and environments will, at least, do no harm.

Can wants be subjected to moral imperatives? We have noted that the problems of the world ecosystem often reduce to ecologically perverse wants. Usually, this simply means that a perfectly reasonable want has gotten out of hand—too many people want the same thing. Recall, however, the perverse wants. Environmentalists have often retooled their wants. This is easy to do; wants are incredibly flexible in the human animal. People may choose to exercise wants for organic produce, ecotourism, efficiency, diversity, creativity, homemade music and crafts, and beautiful scenery.

Enjoyment of nature tends to be a solitary thing—even the sociable Chinese saw it in terms of a loner meditating in the mountains. This makes it a poor fit in this contemporary world, where people are desperate for anything that will bring them into contact with others.

We need a richer, fuller life of contentment and enjoyment, in which individuals are free to find their bliss without being forced to conform for conformity's sake. This was the vision, now almost forgotten, of the great humanists, from Erasmus to Kant. Such a society would be genuinely caring and responsible. It would be environmentally moral without needing to live by a specifically environmental ethic or ecological code. It would be a world where natural resources are used, but used as efficiently as possible, so that the maximum benefits are derived from minimal impact. As Leopold argued, it would be a world of small mixed farms, indigenous land-use systems, selective logging, and compact cities. It would be a world in which consumers preferred quality and variety to quantity. Above all, it would be a world in which people preferred humans to things, and defended themselves by solidarity rather than by hate. The narrow environmentalism that prefers nature to people, or even sees them in competition, is morally and pragmatically untenable.

Ecological ethics in other societies around the world are based on wider concepts of humanity—concepts that are like the psychologists' idea of the total person—an individual with emotional, aesthetic, and spiritual dimensions, as well as rational cognition. Chinese, Navaho, and ancient Greek thinkers agreed that the aesthetic side of human life is inseparably fused with the moral, emotional, and intellectual side. These converge very closely on Levinas's perceptions. The West's loss of such perceptions is one cause of the modern failure of conservation action. We have not recognized how deeply humans need a beautiful environment to be human.

LOVING NATURE

Above all, an environmental ethic has to be grounded, at some remove, in love for people and nature. It has to be real and intense interest, enjoyment, enthusiasm, and warmth for concrete persons and things. Without this neo-Aristotelian virtue (Wensveen 2000), no amount of cold or puritanical morality will do any good for the environment or for anything else. If we do not feel the death of shorebirds and warblers, we will not care until too late. Part of the lesson of the canaries in the mine is that canaries were cherished; if the miners had carried rats, they would not have cared enough at the deaths to take the lesson.

Probably the greatest enemy of such morality today is not simply materialism—bad enough in itself—but technism: obsessive concern for, interest in, and devotion to the most complicated things made by people. Obsession with electronic gadgets and media entertainment is the most obvious form, but dedication to internal combustion vehicles, high-tech weaponry, and legal complexities are all part of the picture. People are naturally more concerned with the human world than with nature, and they extend the human world to the most elaborate things humans create—in today's global village, electronics and politics.

Obsession with these leads to planning the world around them. A world planned for automobiles and planes, and for people who all have computers, cell phones, and iPods, is not a world that can accommodate anything natural. There aren't enough resources to go round. Of course, such a world cannot accommodate the poor, either—the billions who will never afford such amenities. They pay for all, by losing the raw resources and the land; they lose all they have and reap no benefits. Almost no one seems to care that the 225 richest people in the world make more than the 1.7 billion poorest (Bergman 2007).

In so far as Levinas is right, the deepest enemy is indifference. This includes lukewarm love, including lukewarm religion. Religion has been the vehicle for the best defense of the environment, over tens of thousands of years. Today, religion continues its role as leader in environmental morality and defense (Anderson 1996; Tucker and Grim 1994; Woodard 2007). But love for the environment cannot be the fuzzy, generalized, abstract love that too many Christians and other religious and spiritual teachers have discussed and advocated over the millennia. The good is enemy of the best.

Loving nature, however, is obviously not adequate in itself. It has to be combined with doing something to save and protect nature, *and those who depend on it directly for a livelihood.* Yes, that means ranchers and loggers too, but the hard problem today is the billion impoverished rural people who depend on living with, in, and from natural resources. We have to live with the nonhuman world. We have to see everything as people in nature, without separating the realms. Otherwise, we will either plan just for people (and not other organisms), which is absolute suicide in today's desperately food- and fuel-short world, or we will plan to preserve nature at the expense of local people, in which case we will guarantee implacable opposition from them. The separation of people from nature was a mistake in the first place. It is one we can no longer afford.

POSITIVE SOLUTIONS

All this adds up to the most important single message: *We have to look for positive solutions that deal with positive qualities such as enjoyment, love, and social welfare.* The worst thing about public life in the last 40 years has been its increasing negativism. More and more critiques have emerged, with less and less to offer by way of alternatives to the endless list of things they criticize. My field of anthropology used to be based on real interest in people, and was dedicated to understanding them. It is now largely taken up with these sterile critical documents, denouncing multifarious evils, but offering nothing for the future.

No one ever profits from this. Denouncing disease doesn't work; one has to cure it. Denouncing war doesn't work; one has to create conflict resolution mechanisms. Denouncing floods, hurricanes, and droughts won't stop them. If there are no solutions, there will be no improvement.

Without solidarity in the pursuit of pragmatic process goals for a better environment, the world will be lost and humanity will die, along with almost all other species.

FINALLY, A CODE

We can now specify a brief moral code (see Rolston 1988; Norton 1991).

First, the environment is to be kept as rich, diverse, and sustainably productive as possible, consonant with maintaining human life and (at least in scientific reserves) large samples of undisturbed ecosystems. The very minimum of ecological responsibility is that we maintain a habitat we can survive in. Even that is currently not even remotely approached, in much of the world. The most basic goal of individuals and of society should be to increase warm appreciation of people and the environment: love, or at least tolerance, and valuing diversity.

Second, human freedom is to be maximized—explicitly in the sense that one person's rights (including a fair share of rights to all common-pool or partially common-pool resources) extend until they actively interfere with another person's equal rights: "My rights stop where yours start." This attitude also involves basing society more on negotiation and democratic process, and less on the exercise of authority. Responsibility and high personal standards are obviously essential.

Third, this has to happen in a wider context of solidarity. People have to care for each other. Intolerance, bloody-minded defiant aggression, and cold indifference to weaker groups are behaviors we can no longer afford.

Fourth, the search for greater knowledge (of all things, but specifically of ecology and social functioning) must be a basic, all-important part of life.

Fifth, there must be a general search for quality rather than quantity. This is the best source of control on population. People given a choice will raise a few children and invest a great deal in each, rather than raising a lot and investing minimally in each. The same principle plays—or should play—across the board. We should want very few material possessions, but those should consist of the finest quality. This would maximize the consumption of skill, as opposed to the consumption of materials.

Sixth, going beyond such legalistic measures, we would wish to see a spirit of respect and love for the natural world and for sustainable, diverse farming and fishing landscapes. The animistic worldviews of traditional peoples may have peopled the world with unreal fantasy-beings, but they have also taught such love and reverence. We need not go back to what Mircea Eliade called "archaic forms of ecstasy" (Eliade 1962), nor does any religion have a monopoly on ecological wisdom or foolishness. Religion, like other social institutions, grows from negotiation, and must maintain open dialogue if it is not to succumb to extremist and inquisitor factions. Totalitarian religious regimes have proved as ecologically irresponsible as other totalitarianisms, and I fear that the champions of spiritual ecology have not taken enough this issue into account. We need to return to a way of seeing the world, not to a particular sight.

SUMMING UP AND EXTENDING

The overwhelming problem is that basic necessary primary goods are now supplied by giant firms that have fused with government and now wield extreme power. They are beyond accountability. They export their real costs of production, usually in the form of environmental damage, to the rest of us.

The human animal is constituted in such a way that worries about far-future or remote situations are not compelling. People worry about the immediate loss of what they love and value. Therefore, loving nature and concern over its demise are the only motives that will unite us in time to save the world.

Loving nature remains the basic need, but is perhaps beyond many in this world. We can be satisfied with some love and much responsibility. The latter is often a more direct spur to action than love, in any case. We have to be responsible toward the environment, but perhaps more toward each other. Solidarity in defending the future of humanity is the major need.

Fear for one's life, one's loved ones, and society in general is a close second among emotional motives. The world environmental crisis is now a deadly reality for a fast-increasing proportion of humanity, and will very soon overwhelm us all.

Health and recreation, and Teddy Roosevelt-type toughness and wilderness values, are not unworthy of attention. Getting outdoors is the last best cure for obesity.

All this will require some sort of totalizing vision—if not religious, at least moral.

Besides the values noted above, tolerance, reasonableness, and good sense seem critical. Science and learning are obviously essential (as shown by the fervor with which certain giant primary-production firms oppose *inconvenient truth*). More generally, long-term, wide-flung horizons remain basic. Economic growth, perhaps paradoxically, is also necessary; we can do it while improving the environment, simply by being efficient.

Against this play the short-term, narrow values of the throughput-maximizing world: progress as destruction, unnaturalism, irresponsibility, and above all, prejudice and hatred. Passive conformity has proved disastrous also. Obviously, corruption and unfair politics, lack of transparency, lack of accountability, and general disunion are no longer practices we can afford.

The most immediate and pragmatic need is serious change and then investment in education. Following that come research and development of better technologies, and some major turnaround on public lands and land management.

More generally, perhaps the best we can hope for is that people will discover responsibility and minimal self-preservation awareness before it is too late. For the farther future, we may yet recover the broader visions of the world that endure in traditional societies: the world as a beautiful, powerful, living place, a realm of spirit and intensity rather than mere stuff to use and throw away. A vision like this allows us to think beyond the throughput-maximizing world.

IS ANY OF THIS PRACTICAL?

Ideally, we would all return to an intimate concern with the nonhuman world, the sort of concern found in traditional societies. But we of the modern world are inevitably committed to technology, and are also heirs to the broadly antinature, antinatural worldview propagated by the Roman Empire and its heirs. The Roman Empire had checks, balances, and qualifications to this view, but still exterminated everything from European lions to North African elephants, and even the silphium plant, a Libyan medicinal herb (Koerper and Kolls 1999). The modern global economy, concerned with machines and throughput rather than sustainability and efficiency, is an extreme derivative.

We can, however, hope for responsibility. Humans can and do care (sometimes, at least) about the survival and economic well-being of their fellow humans. As environmental damage causes more and more deaths and ruined lives, people around the world come to protest and to act, even if they have no special love for nature.

A strong sense of responsibility about taking care of the environment is now astonishingly widespread around the world. I have seen it come to the areas in which I do most of my research: Mexico and China. Neither was especially noted for environmental care 40 years ago, and indeed the nations had strong ideologies of *progress* that involved destroying the natural and substituting the most artificial environment possible. This was an old Spanish legacy in Mexico; in China it was new, deriving from Mao's "struggle against nature" slogans. Countertraditions—indigenous cultures in Mexico, traditional attitudes in China—survived, but had little nationwide effect. Today the situation is far different, and almost all young people are aware that they should be responsible about the environment. Practice inevitably fails to come up to ideals, but the trajectory is clear and hopeful.

The absolutely critical point—the one argued most strongly in the present book—is that this responsibility has to add up to a strong sense of unity. The opposition is united: Powerful primary-production interests stand together, and their right-wing backers are famous for loyalty and solidarity. However much they may fight among themselves, they always present a united front against progressives of all stripes. The progressives and liberals, however, are famously divided, and keep losing close elections because of breakaway movements. We remember how Gore and the environmental cause were ruined by Ralph Nader in 2000; fewer remember that Nixon defeated Humphrey in 1968 because of critical defections to the Peace and Freedom Party, especially in California. The right continues to appeal successfully to racism and religious bigotry, and to exploit political squabbles on the center and left. The world cannot afford this any longer.

Loving nature, however, remains essential, and thus environmental education with real field experience is probably the most important single thing to do now. More broadly, love for nature has to be culturally constructed—if not in religion, in public ideology and awareness—to have any effect.

ENDPIECE

Ecotopia would be a place of love and freedom. It would be a world of free choices and many options, rather than a world of uniformity. Ecotopia, like sustainability, is a process goal.

Without a vision of a better future, solving specific problems will not work. We will never know which problems are most urgent, or what strategy to take if we wish to improve the world rather than just slow its deterioration.

No community can exist, let alone manage resources, without some degree of solidarity. No real community can be held together by mere force or terror. A prison or slave camp may be held together in such a way, and even a whole country for a time, but such places do not function as self-sustaining communities.

People must want a good environment. They have to love it. Only this will motivate the necessary moral commitment and actual political and economic behavior. Minimally, people must want a safe and relatively unpolluted environment, and must realize that this would help, not hurt, the economy. Maximally, people will want to watch birds, identify wildflowers, and in every way take some real pleasure not only in seeing the environment but also in learning about its complexity. People will want to eat a varied diet, not just white flour, vegetable oil, beef, and sugar.

The real thrust of the environmental movement, then, should be toward producing and maintaining, through deliberate human action, a livable environment in those areas where humans live. We need to plant trees, restore marshes, bring fish back to rivers, bring sustainable management techniques back into farming, and work in our own gardens. Simply to save the livelihood of our children and grandchildren, we have to work to reverse the forces of degradation and destruction, and create anew, in the settled lands, the diversity, the human scale, the sustainable management techniques, and the beauty that we have lost. Daniel Janzen (1998) is correct in recommending that we see the world as one big garden, to be cared for as such. In Judeo-Christian tradition, Adam and Eve were charged by God to care for the Garden of Eden. Every home garden should be a microcosm of the world we want.

Voltaire's Candide and Pangloss ended their philosophical journeys through the world by cultivating their garden. We can do no better than follow their example.

I will, however, go further, and in the terms of my Calvinist background, speak of "the priesthood of all believers"—but for "believers," substitute "beings."

Notes

CHAPTER 1

1. It appears that this image is far enough from modern experience to need some explanation: I recently read an alarmist text, which I will charitably leave anonymous here, that said, "The canaries are singing loudly." The phrase actually implies quite the opposite image. It comes from old mining days. Canaries are very sensitive to mine gases, and succumb when the gases get dangerous. Humans are more resistant. Thus, miners would carry canaries with them into the mines. When the canaries keeled over, the miners promptly got out of there.

2. I would humbly submit that the ecological and environmental movements are not wholly immune to this mindset.

3. The best of these surveys to come under my gaze is that of Kempton, Boster, and Hartley (1995), which reviews previous ones.

4. See Ponting (1991). On the Maya, see especially the formidable case made for ecological devastation at Copan (Sharer and Traxler 2006). From my own research, I find it impossible to believe that the population densities in classic Maya times in my research area—western Quintana Roo—were sustainable by any imaginable technology, ancient or modern. On the other hand, the population densities Sharer cites are, I believe, exaggerated. Thus, there is still room for debate: Did the Maya overshoot ecologically (Diamond 2005), or engage in too much war (Demarest 2004), or simply succumb to drought (Gill 2000), or decline for other reasons, not all known (Webster 2002)? For China, the classic work of Mallory (1922) has not been surpassed, but see also Marks (1998); for Polynesia, see Kirch (1994, 1997).

5. At this point I will reiterate a credo that I feel I must put in all my books some-where: I am a behavioral scientist, and to me science is a search for ever more accu-rate understanding of the world—not a list of "facts" or a socially constructed,

culturally specific game. There are, to be sure, scientists who do seem to "list facts," and others, perhaps better called "scientists" with scare quotes, who disguise political agendas under a coating of factoids. A good example of the latter class is *The Bell Curve* (Herrnstein and Murray 1994), a book exactly as scientific as a newspaper astrology column, in spite of its portentous statistical trappings. But, *pace* the postmodernists, not all or even most scientists act like this. Most of us patiently chip away at the edges of ignorance, making tentative statements and then trying our best to support them with more and more conclusive observations—or to abandon them when the observations go the other way. I shall develop this below, especially that last clause.

6. In addition to E. P. Thompson (1963) and Endre Nyerges (1997), my inspirations in my interactive, practice-oriented approach to social science have been Pierre Bourdieu (1978, 1990), George Homans (1961), and Jean Lave (1988). I have also profited much from Jurgen Habermas's theories of civil society (Habermas 1984) and from the many scholars who have studied the creation of political forms and entities, such as Benedict Anderson (1991) and James Scott (1998).

7. I have noted elsewhere (Anderson 1996) that general human needs include not only food, clothing, and shelter, but also social acceptance and a sense of control over one's life. Most people feel, also, a genuine need for beauty, peace, and other aesthetic goods. Economists often write as if people wanted nothing but money; this is, of course, not the case. Money is merely a way to buy what is really wanted, and, as we all know, "money can't buy everything."

One of the problems with Marxism was that Marx, though aware of the complex needs of the human animal and the need for social morality, really did believe that material wealth came first and that economic reorganization would solve all the problems. The experiment was tried, and it failed.

CHAPTER 4

1. I can certainly imagine many other scenarios, and maybe they would be better than mine. Create your own, dear reader. In the words of a graffito I once saw painted on a Berkeley newsstand: "If you don't like the news, make your own!"

CHAPTER 5

1. A useful review of spectacular regulatory and justice failures in the 20th century (Harremoës et al. 2002), from fisheries management and asbestos pollution to PCBs and England's "mad cow disease" crisis, leads to considerable refinement of the precautionary principle. The precautionary principle certainly lies at the heart of much of environmental justice (Shrader-Frechette 1991, 2002). "The PP" is making slow but real progress as a way to evaluate projects and practices that impact the environment.

General sources bring some of these considerations together (Adamson et al. 2002; Agyeman et al. 2003; Turner and Wu 2002). Latin America has not lagged behind, as shown by a major collection edited by a world leader in human ecology and environmentalism, Enrique Leff (2001).

The University of California, Berkeley's Workshop on Environmental Politics has issued a 135-page annotated bibliography, *Environmental Justice and Environmental Racism* (Turner and Wu 2002), that covers most relevant areas.

Finally, the environmental justice movement has matured enough to be routinely shortened to "EJ" and to have elicited a recent "critical appraisal" that expresses concerns over "stagnation" (Benford 2005) and makes calls to renewed zeal (Pellow and Brulle 2005; the chapters in this edited volume are highly programmatic).

A related literature on violence, especially the experiences of victims and the ideologies of committers (Das and Kleinman 2000; Feldman 1991), is highly relevant, even when it does not discuss the environment per se. Environmental destruction has been used as a part of genocide in the past. The most famous case in American history was the extermination of the buffalo by white hunters on the American plains, done partly to starve the Plains Indians into submission or death. Irish history reveals episodes of deforestation by the English (especially in the 17th century) to weaken Irish resistance and deny Irish guerrillas a hiding place. Such scorched-earth practices are worldwide. They are universal in war and routinely carry over into peacetime as ways to control or exterminate targeted populations.

Several excellent studies that bring out the complexity of political-environmental discourse are found in Beinart and McGregor (2003), Carrier (2004), and Zimmerer and Bassett (2003). Dealing directly with ecological, economic, and political issues is necessary. If this is kept in mind, discourse analysis can be valuable, especially to remove the smoke screen of carefully calculated "discourse" that often cloaks environmental injustice (Markowitz and Rosner 2002; Stauber and Rampton 1995).

Other major studies of the human rights issues involved can be found in several books (Dobson 1990, 1998, 1999; Picolotti and Taillant 2003). Dobson, in particular, has provided dense and closely argued tracts on the interface between concepts of justice in general (such as those of John Rawls [v. Rawls 1971], Robert Nozick, and Michael Walzer), sustainability (ever hard to define), and environmental justice. K. Shrader-Frechette (1991, 2002) has devoted a career to the closely related issue of risk—how to evaluate environmental risks from an ethical point of view.

For much of these references and other help with this chapter, I thank Yolanda Moses, who started me on this project. Others who have been especially helpful in preparing this report include Rebecca Austin, Barbara Brower, Barbara Rose Johnston, Devon Peña, Richard Peterson, and Terre Satterfield.

2. Hong Kong under the British is said to be the nearest to a free market that the world has known. I spent two years researching Hong Kong's ecology, economy, and society, and observed a great deal of environmental damage as well as much good environmental management. I never could decide whether the damage was due more to the market freedom or to the remaining market imperfections. Others seem equally puzzled.

CHAPTER 6

1. The world's dozen largest corporations, as of 2007, are, in order: Wal-Mart; Exxon Mobil, the denier of global warming and funder of most anti-global-warming propaganda; Royal Dutch Shell; BP; General Motors (still the top in spite of chronic troubles); Toyota; Chevron; DaimlerChrysler; ConocoPhillips; General Electric; Ford. In other words, big oil and big cars, plus one retailer. Other oil companies are close; China National Petroleum is 24th, Indonesia's oil 25th, Mexico's Pemex 34th. The rest of the world's leading oil and automobile corporations are all in the top 100, along with a miscellany of banks, insurance firms, retailers, and conglomerates. No agribusiness or mining firm makes it, but several still manage to be hugely powerful in many countries.

Common Property Management and Related Issues

As in the large, so in the small: The downfall of investigative reporting in the *Los Angeles Times* and neighboring newspapers is felt in local politics as well. County supervisors in Riverside and San Bernardino Counties, where I once lived, are largely from the development sector. They routinely grant variances to large developers, exempting them from state and county environmental regulations. Large campaign donations and other perquisites are naturally involved. Ordinary people may have to go through endless bureaucratic hoops to add a room to a house, but giant developers build in active floodplains without hindrance. (One is doing so in Lytle Creek Wash, San Bernardino County, as I write, in an area that sees walls of water carrying bushel-sized boulders in every heavy rainy period. No flood control work is being done.) In past years, newspapers carried stories on these matters, and public protests stopped the most flagrant abuses. Since the local papers have been taken over by vast right-wing combines, news has been reduced up to 70 percent or more and environmental coverage essentially eliminated. No reports or protests now ensue. If public hearings are actually held and public protest surfaces, the supervisors or other agencies simply postpone the hearings over and over. The developers pay their representatives, whose job includes going to hearings—and who are, in any case, given to understand when a postponement is likely. The citizen protesters have to sacrifice work time or other important times in order to attend. Protest soon dies.

Again, the problem is not with government. It has the laws and is charged by these laws with stopping such developments. Nor is the problem with private enterprise; free competition and a free press would stop the abuses. The problem is the linkage between certain privileged firms and their governmental sponsors.

2. Of course, the cruder forms of "libertarianism," like the cruder forms of socialism, are plain silly. One cannot take seriously the preposterous excesses of "American capitalism" as seen in, for example, the writings of Richard Pombo (Pombo and Farah 1996), a rancher and real estate developer who opposes all regulation of private property and all aid to the poor, but who receives, with enthusiasm, the huge subsidies to ranching passed out by the U.S. Government. In fact, he was also a U.S. Congressman, with a solid record of opposition to the Endangered Species Act, the

Environmental Protection Agency, and all interference with private property—even interference necessary to prevent direct and serious damage to other people. (Let 'em suffer, or buy their way out of it.) Pombo also compiled a solid record of voting against civil rights and minority rights. Of course, genuine free market environmentalists argue that Pombo is not a capitalist but a government servant who uses his position to steer vast amounts of taxpayers' money to his own line of resource depletion. He was defeated for re-election in 2006.

References

Abrahamsen, Rita. 2000. Disciplining Democracy: Development Discourse and Good Governance in Africa. London: Zed Books.

Acheson, James M. 1998. "Lobster Trap Limits: A Solution to a Communal Action Problem." Human Organization 57:43–62.

———. "Lobster and Groundfish Management in the Gulf of Maine: A Rational Choice Perspective." Human Organization 65:240–252.

Adamson, J.; Mei Mei Evans; Rachel Stein (eds.). 2002. The Environmental Justice Reader: Politics, Poetics, and Pedagogy. Tucson: University of Arizona Press.

Agrawal, Arun. 2005. "Environmentality: Community, Intimate Government, and the Making of Environmental Subjects in Kumaon, India." Current Anthropology 46:161–190.

———. 2006. Environmentality: Technologies of Government and the Making of Subjects. Durham: Duke University Press.

Ainslie, George. 1993. Picoeconomics: The Strategic Interaction of Successive Motivational States with the Person. Cambridge: Cambridge University Press.

Agyeman, Julian; Robert Bullard; Bob Evans. 2003. Just Sustainabilities: Development in an Unequal World. London: Earthscan.

Alvard, Michael. 1995. "Interspecific Prey Choice by Amazonian Hunters." Current Anthropology 36:789–818.

Anderson, Benedict. 1991. Imagined Communities. 2nd edn. London: Verso.

Anderson, E. N. 1987. "A Malaysian Tragedy of the Commons." In The Question of the Commons, Bonnie McCay and James Acheson (eds.). Tucson: University of Arizona Press. Pp. 327–343.

———. 1996. Ecologies of the Heart. New York: Oxford University Press.

———. 1999 (orig. 1969). "The Life and Culture of Ecotopia." In Reinventing Anthropology, Dell Hymes (ed.). New edn. Ann Arbor: University of Michigan Press. Pp. 264–283.

———. 2005. Political Ecology in a Yucatec Maya Community. Tucson: University of Arizona Press.

———. ms 1. The Morality of Ethnobiology. Available from author.

———. ms 2. Madagascar on My Mind. Available from author; earlier version posted on website, www.krazykioti.com.

Anderson, E. N., and Marja L. Anderson. 1978. Fishing in Troubled Waters. Taipei: Orient Cultural Service.

Anderson, E. N., and Christopher Chase-Dunn (eds.). 2005. "The Rise and Fall of Great Powers." In The Historical Evolution of World-Systems. Christopher Chase-Dunn and E. N. Anderson (eds.). New York: Palgrave MacMillan. Pp. 1–19.

Anderson, E. N., and Felix Medina Tzuc. 2005. Animals and the Maya in Southeast Mexico. Tucson: University of Arizona Press.

Anderson, M. Kat. 2005. Tending the Wild: Native American Knowledge and the Management of California's Natural Resources. Berkeley: University of California Press.

Anderson, Terry, and Donald Leal. 1997. Enviro-Capitalists: Doing Good while Doing Well. Lanham, MD: Rowman & Littlefield Publishers.

———. 1991. Free Market Environmentalism. San Francisco: Pacific Research Institute for Public Policy and Westview.

Anderson, Terry, and Randy Simmons (eds.). 1993. The Political Economy of Customs and Culture: Informal Solutions to the Commons Problem. Lanham, MD: Rowman and Littlefield.

Arax, Mark. 1997. "Flooding Doesn't Slow Plans to Build on Riverbed." Los Angeles Times, Jan. 13, pp. 1, 16.

Ascher, William. 1999. How Governments Waste Natural Resources. Baltimore: Johns Hopkins University Press.

Attwood, Bain, and Andrew Marks (eds.). 1999. The Struggle for Aboriginal Rights: A Documentary History. Crows Nest, NSW: Allen and Unwin.

Aumeeruddy, Yildiz. 1994. Local Representations and Management of Agroforests on the Periphery of Kerinci Seblat National Park, Sumatra, Indonesia. Paris: UNESCO. People and Plants Working Paper #3.

Baden, John, and Douglas Noonan (eds.). 1998. Managing the Commons. Bloomington, IN: Indiana University.

Baden, John, and Donald Snow (eds.). 1997. The Next West. Washington: Island Press.

Balter, Michael. 2008. "Why We're Different: Probing the Gap between Apes and Humans." Science 319:404–405.

Bandura, Albert. 1982. "Self-Efficacy Mechanism in Human Agency." American Psychologist 37:122–147.

———. 1986. Social Foundations of Thought and Action. Englewood Cliffs, NJ: Prentice-Hall.

Barbour, Haley. 1996. Agenda for America: A Republican Direction for the Future. Washington: Regnery.

Barlett, Donald L., and James B. Steele. 1998. "Paying a Price for Polluters: Many of America's Largest Companies Foul the Environment but Clean up on Billions of Dollars in Tax Benefits." Time, Nov. 23, 1998, pp. 72–80; "Sweet Deal," ibid., pp. 81–82.

Barrett, Scott. 2003. Environment and Statecraft: The Strategy of Environmental Treaty-Making. OUP.

Barry, John. 1999. Green Politics. London: Routledge.

Baskin, Yvonne. 1997. The Work of Nature: How the Diversity of Life Sustains Us. Washington, DC and Covelo, CA: Island Press.

Bates, Daniel G. 2005. Human Adaptive Strategies: Ecology, Culture, and Politics. 3rd edn. Boston: Pearson Allyn and Bacon.

Beinart, William, and Joann McGregor. 2003. Social History and African Environments. Oxford: James Currey in collaboration with Ohio University Press.

Bender, Barbara, and Margot Winer (eds.). 2001. Contested Landscapes: Movement, Exile and Place. Oxford: Berg.

Benford, Robert. 2005. "The Half-life of the Environmental Justice Frame: Innovation, Diffusion, and Stagnation." *In* Power, Justice, and the Environment: A Critical Appraisal of the Environmental Justice Movement, David Naguid Pellow and Robert J. Brulle (eds.). Pp. 37–54. Cambridge, MA: MIT Press.

Bennett, John W. 1976. The Ecological Transition: Cultural Anthropology and Human Adaptation. New York: Academic Press.

Bennett, W. Lance; Regina Lawrence; Steve Livingston. 2007. When the Press Fails: Political Power and the News Media from Iraq to Katrina. Chicago: University of Chicago Press.

Bergman, Justin. 2007. "Organized Crime: $2 Trillion Last Year, Study Says." Seattle Times, Sept. 11, p. A9, from Associated Press.

Berkes, Fikret. 1999. Sacred Ecology: Traditional Ecological Knowledge and Resource Management. Philadelphia: Taylor and Francis.

Berkes, Fikret; Johan Colding; Carl Folke. 2000. "Rediscovery of Traditional Ecological Knowledge as Adaptive Management." Ecological Applications 10:1251–1262.

Berman, Marc; John Jonides; Stephen Kaplan. 2008. "The Cognitive Benefits of Interacting with Nature." Psychological Science 19:1207–1212.

Biello, David. 2008. "Not Going Out to Play." Scientific American, April, p. 36.

Billing, Jennifer, and Paul W. Sherman. 1998. "Antimicrobial Functions of Species: Why Some Like It Hot." Quarterly Review of Biology 73:3–49.

Blackburn, Thomas, and Kat Anderson (eds.). 1993. Before the Wilderness. Socorro, NM: Ballena Press.

Böge, Völker. 1999. "Mining, Environmental Degradation and War: The Bougainville Case." Suliman 1999:211–227.

Bookchin, Murray. 1982. The Ecology of Freedom. Palo Alto: Cheshire.

Borgerhoff Mulder, Monique, and Peter Coppolillo. 2005. Conservation: Linking Ecology, Economics and Culture. Princeton: Princeton University Press.

Botkin, Daniel. 1990. Discordant Harmonies: A New Ecology for the Twenty-First Century. New York: Oxford University Press.

Bourdieu, Pierre. 1978. Outline of a Theory of Practice. Cambridge: Cambridge University Press.

———. 1990. The Logic of Practice. Stanford: Stanford University Press.

Bovard, James. 1991. The Farm Fiasco. San Francisco: Institute of Contemporary Studies.

Brandt, Richard B. 1954. Hopi Ethics: A Theoretical Analysis. Chicago: University of Chicago Press.

Brechin, Steven R.; Peter R. Wilshusen; Crystal L. Fortwangler; Patrick C. West (eds.). 2003. Contested Nature: Promoting International Biodiversity with Social Justice in the Twenty-First Century. Albany: SUNY.

Broome, John. 2008. "The Ethics of Global Warming." Scientific American, June, pp. 97–102.

Brown, Lester (ed.). 1999. State of the World 1999. Washington, DC: Worldwatch Institute.

———— (ed.). 1999. Vital Signs 1999. Washington, DC: Worldwatch Institute.

————. 1995. Who Will Feed China? Wake-up Call for a Small Planet. New York: W. W. Norton.

Brown, Michael F. 2003. Who Owns Native Culture? Cambridge, MA: Harvard University Press.

Brunnschweiler, C. N., and E. H. Bulte. 2008. "Linking Natural Resources to Slow Growth and More Conflict." Science 320:616–618.

Bryant, Bunyan (ed.). 1995. Environmental Justice: Issues, Policies and Solutions. Washington, DC: Island Press.

Bryce, George. 1968 (orig. 1904). The Hudson's Bay Company. New York: Burt Franklin.

Buber, Martin. 1947. Between Man and Man. London: Routledge and Kegan Paul.

————. 1991. Tales of the Hasidim. Olga Marx (tr.). New York: Schocken Books.

Buchowski, Michal; David Kronenfeld; William Peterman; Lynn Thomas. 1994. "Language, Nineteen Eighty-Four, and 1989." Language in Society 23:555–578.

Bullard, Robert. 1990. Dumping in Dixie: Race, Class, and Environmental Quality. Boulder, CO: Westview.

———— (ed.). 1994. Unequal Protection: Environmental Justice and Communities of Color. San Francisco: Sierra Club Books.

———— (ed.). 2005. The Quest for Environmental Justice: Human Rights and the Politics of Pollution. Fwd by Maxine Waters. Sierra Club 2005

Bullard, Robert; Glenn S. Johnson; Angel O. Torres. 2000. Race, Equity, and Smart Growth: Why People of Color Must Speak for Themselves. Environmental Justice Resource Center at Clark Atlanta University.

Bunker, Stephen G., and Paul S. Ciccantell. 2005. Globalization and the Race for Resources. Baltimore: Johns Hopkins University Press.

Burnett, John. 1966. Plenty and Want. Harmondsworth, Sussex: Penguin.

Butler, Declan. 2007. "Darfur's Climate Roots Challenged." Nature 447:1038.

Callenbach, Ernest. 1975. Ecotopia. Berkeley: Banyan Tree Books.

Callicott, J. Baird. 1989. "American Indian Land Wisdom? Sorting out the Issues." Journal of Forest History, Jan., 35–42.

————. 1990 "Genesis and John Muir." ReVision 12:31–47.

———. 1993. "The Conceptual Foundations of the Land Ethic." *In* Environmental Philosophy: From Animal Rights to Radical Ecology (2nd edn.). Michael E. Zimmerman, J. Baird Callicott, George Sessions, Karen J. Warren, John Clark, eds. Pp. 110–134. New York: Prentice-Hall.

———. 1994. Earth's Insights. Berkeley: University of California Press.

———. 1995. "Animal Liberation: A Triangular Affair." *In* Environmental Ethics. Robert Elliot (ed.). Pp. 29–59. Oxford: Oxford University Press.

Callicott, J. Baird, and Michael P. Nelson (eds.). 1998. The Great New Wilderness Debate: An Expansive Collection of Writings Defining Wilderness from John Muir to Gary Snyder. Athens: University of Georgia Press.

Caplan, Bryan. 2007. The Myth of the Rational Voter. Princeton: Princeton University Press.

Carrier, James G. (ed.). 2004. Confronting Environments: Local Understanding in a Globalizing World. Walnut Creek, CA: AltaMira.

Carter, Alan. 1999. A Radical Green Political Theory. London: Routledge.

Checker, Melissa. 2005. Polluted Promises: Environmental Racism and the Search for Justice in a Southern Town. New York: New York University Press.

Cheney, Dorothy, and Robert M. Seyfarth. 2007. Baboon Metaphysics. Chicago: University of Chicago Press.

Chivian, Eric, and Aaron Bernstein. 2008. Sustaining Life: How Human Health Depends on Biodiversity. New York: Oxford University Press.

Choi, Charles Q. 2006. "Voting with the Heart." Scientific American, Dec., 36–38.

Clark, Kate, and Alan O'Connor. 2008. "Dead Sea." Business and Finance [Ireland], June 8, pp. 68–69.

Cleveland, David, and Stephen Murray. 1997. "The World's Crop Genetic Resources and the Rights of Indigenous Farmers." Current Anthropology 38:477–515.

Colding, Johan, and Carl Folke. 2001. "Social Taboos: 'Invisible' Systems of Local Resource Management and Biological Conservation." Ecological Applications 11:584–600.

Cole, Luke W., and S. R. Foster. 2001. From the Ground Up: Environmental Racism and the Rise of the Environmental Justice Movement. New York: New York University Press.

Collier, Paul. 2007. The Bottom Billion: Why the Poorest Countries Are Failing and What Can Be Done About It. New York: Oxford University Press.

Collins, Randall. 2001. Interaction Ritual Chains. Princeton: Princeton University Press.

Comaroff, Jean. 1985. Body of Power, Spirit of Resistance: The Culture and History of a South African People. Chicago: University of Chicago Press.

Cooney, Rosie, and Barney Dickson (eds.). 2005. Biodiversity and the Precautionary Principle: Risk and Uncertainty in Conservation and Sustainable Use. London: Earthscan.

Daily, Gretchen (ed.). 1997. Nature's Services: Societal Dependence on Natural Ecosystems. Washington, DC, and Covelo, CA: Island Press.

Daly, Herman, and John B. Cobb Jr. 1994. For the Common Good: Redirecting the Economy toward Community, the Environment, and a Sustainable Future. 2nd edn. Boston: Beacon.

Damasio, Antonio. 1994. Descartes' Error. New York: Putnam.

Das, Veena, and Arthur Kleinman (eds.). 2000. Violence and Subjectivity. Berkeley: University of California Press.

Davis, Thomas. 2000. Sustaining the Forest, the People, and the Spirit. Albany: SUNY Press.

Dawkins, Richard. 1976. The Selfish Gene. Oxford: Oxford University Press.

Deane-Drummond, Celia. 2006. "Environmental Justice and the Economy: A Christian Theologian's View." Ecothology 11:294–310.

De Waal, Frans. 1996. Good Natured. Cambridge, MA: Harvard University Press.

———. 2005. Our Inner Ape. New York: Riverhead Books (Penguin Group).

Delcourt, Paul, and Hazel Delcourt. 2004. Prehistoric Native Americans and Ecological Change: Human Ecosystems in Native North America since the Pleistocene. New York: Cambridge University Press.

Demarest, Arthur. 2004. Ancient Maya: The Rise and Fall of a Rainforest Civilization. Cambridge: Cambridge University Press.

Delong, Claudio. 2005. "The Political Ecology of Deforestation in Thailand." Geography 90:225–237.

D'Estrée, Tamra Pearson, and Bonnie G. Colby. 2004. Braving the Currents: Evaluating Environmental Conflict Resolution in the River Basins of the American West. Boston: Kluwer Academic Publications.

Deur, Douglas, and Nancy J. Turner (eds.). 2005. Keeping It Living: Traditions of Plant Use and Cultivation on the Northwest Coast of North America. Seattle: University of Washington Press; Vancouver: University of British Columbia Press.

Diamond, Jared. 2005. Collapse: How Societies Choose to Fail or Succeed. New York: Viking.

Dichter, Thomas W. 2003. Despite Good Intentions: Why Development Assistance to the Third World Has Failed. Amherst & Boston: University of Massachusetts Press.

Dionne, E. J., Jr. 2008. Sould Out: Reclaiming Faith and Politics after the Religious Right. Princeton: Princeton University Press.

Dobson, Andrew. 1999. Justice and the Environment: Conceptions of Environmental Sustainability and Theories of Distributive Justice. New York: Oxford University Press.

——— (ed.). 1999. Fairness and Futurity: Essays on Environmental Sustainability and Social Justice. Oxford: Oxford University Press.

Dombrowski, Kirk. 2002. "The Praxis of Indigenism and Alaska Native Timber Politics." American Anthropologist 104:1062–1073.

Domke, David, and Kevin Coe. 2007. The God Strategy: How Religion Became a Political Weapon in America. New York: Oxford University Press.

Dore, Muhammad, and Timothy Mount. 1999. Global Environmental Economics: Equity and the Limits of Markets. Oxford: Blackwell.

Douglas, Mary. 1966. Purity and Danger. New York: Praeger.

Downs, Anthony. 1957. An Economic Theory of Democracy. New York: Harper and Row.

Doyle, Rodger. 1998. "How Congress Voted on the Environment." Scientific American, Nov., pp. 36–37.

Drahos, Peter. 2002. Information Feudalism. New York: New Press.

Dubos, Rene. 1959. Mirage of Health. New York: Harper.

Durkheim, Emile. 1933. The Division of Labor in Society. New York: Free Press.

———. 1995 [1912]. The Elementary Forms of Religious Life. Karen E. Fields (tr.). New York: Free Press.

Echeverria, John, and Raymond Booth Ely (eds.). 1995. Let the People Judge: Wise Use and the Private Property Rights Movement. Washington, DC, and Covelo, CA: Island Press.

Economy, Elizabeth C. 2005. The River Runs Black: The Environmental Challenge to China's Future. Ithaca, NY: Cornell University Press.

Edmonds, Richard Louis (ed.). 1998. Managing the Chinese Environment. Oxford: Oxford University Press.

Ehrlich, Paul, and Anne Ehrlich. 1968. The Population Bomb. New York: Ballantine.

———. 1972. Population, Resources, Environment. San Francisco: W. H. Freeman.

———. 1996. Betrayal of Science and Reason. Washington, DC, and Covelo, CA: Island Press.

Eichenwald, Kurt. 2000. The Informant. New York: Broadway Books.

Eliade, Mircea. 1962. Shamanism: Archaic Techniques of Ecstasy. New York: Pantheon.

Elliott, Larry, and Dan Atkinson. 2008. The Gods That Failed. London: Bodley Head.

Elster, Jon. 1989. Nuts and Bolts for the Social Sciences. Cambridge: Cambridge University Press.

———. 1993. Political Psychology. Cambridge: Cambridge University Press.

Erraro, Paul J. and Agnes Kiss. 2002. "Direct Payments to Conserve Biodiversity." Science 298:1718–1719.

Escobar, Arturo. 1999. "After Nature: Steps to an Antiessentialist Political Ecology." Current Anthropology 40:1–30.

Evans, L. 1998. Feeding the Ten Billion: Plants and Population Growth. Cambridge: Cambridge University Press.

Faber, Daniel. 1998. The Struggle for Ecological Democracy: Environmental Justice Movements in the United States. New York: Guilford Press.

Fairhead, James, and Melissa Leach. 1996. Misreading the African Landscape: Society and Ecology in the Forest-Savanna Mosaic. African Studies Series, 90. UC.

Farnsworth, Norman, and D. Soejarto. 1985. "Potential Consequences of Plant Extinction in the United States on the Current and Future Availability of Prescription Drugs." Economic Botany 39:231–240.

Faust, Betty B; E. N. Anderson; John G. Frazier (eds.). 2004. Rights, Resources, Culture, and Conservation in the Land of the Maya. Westport, CT: Greenwood.

Fei, John, and Gustav Ranis. 1997. Growth and Development from an Evolutionary Perspective. Oxford: Blackwell.

Feldman, Allen. 1991. Formations of Violence: The Narrative of the Body and Political Terror in Northern Ireland. Chicago: University of Chicago Press.

Feng Liu. 2001. Environmental Justice Analysis: Theories, Methods, and Practice. Boca Raton, FL: Lewis Publishers.

Feshbach, Murray, and Alfred Friendly Jr. 1992. Ecocide in the USSR: Health and Nature under Siege. New York: Basic Books.

Field, Barry C. 1994. Environmental Economics: An Introduction. New York: McGraw-Hill.

Firth, Raymond. 1959. Social Change in Tikopia: Re-study of a Polynesian Community after a Generation. New York: MacMillan.

Fischetti, Mark. 2006. "Protecting New Orleans." Scientific American, Feb., pp. 65–71.

Fletcher, Martin. 2008. "Thatcher's Son Was Part of Group Behind African Coup." Irish Independent, June 17, p. 36.

Foster, John Bellamy. 1998. "The Limits of Environmentalism without Class: Lessons from the Ancient Forest Struggle in the Pacific Northwest." In The Struggle for Ecological Democracy: Environmental Justice Movements in the United States. Daniel Faber (ed.). Pp. 188–217. New York: Guilford Press.

Foucault, Michel. 1966. Madness and Civilization. New York: Mentor.

———. 1980. Power/Knowledge: Selected Interviews and Other Writings 1972–1977. Colin Gordon (ed.), Colin Gordon, et al (tr.). New York: Pantheon Books.

———. 1990. The History of Sexuality: An Introduction. New York: Vintage.

———. 1997. The Politics of Truth. New York: Semiotext(e).

———. 2001. Fearless Speech. Joseph Pearson, (ed.). Los Angeles: Semiotext(e).

Fowler, Cynthia. 2007. Herpetological and Ethnobiological Knowledge in Vietnam's Cat Tien Biosphere Reserve. Presentation, Society of Ethnobiology, annual meeting, Berkeley, CA.

Frank, Robert. 1988. Passions Within Reason: The Strategic Role of the Emotions. New York: W. W. Norton.

Frank, Thomas. 2004. What's the Matter with Kansas? How Conservatives Won the Heart of America. New York: Metropolitan Books (Henry Holt & Co.).

Fratkin, Elliot. 2004. Ariaal Pastoralists of Kenya. 2nd edn. Boston: Allyn and Bacon.

Friedman, Benjamin. 2005. The Moral Consequences of Economic Growth. New York: Knopf.

Friedman, David. 1999. "The Jackpot Economy." Los Angeles Times, p. B5.

Friedman, Thomas. 2008. "The Democratic Recession." International Herald Tribune, May 3.

Gaffney, Mason, and Fred Harrison (eds.). 1994. The Corruption of Economics. London: Shepheard-Walwyn in association with the Centre for Incentive Taxation Ltd.

Gay, Frank. 1999. "Reducing Cat Predation on Wildlife." Outdoor California, May–June, pp. 4–8.

Gedicks, Al. 2001. Resource Rebels: Native Challenges to Mining and Oil Corporations. Boston: South End Press.

Gedney, N.; P. M. Cox; R. A. Betts; O. Boucher; C. Huntingford; P. A. Stott. 2006. "Detection of a Direct Carbon Dioxide Effect in Continental River Runoff Records." *In* Nature 439: 835–838.

Genovese, Eugene. 1974. Roll, Jordan, Roll: The World the Slaves Made. New York: Vintage.

Gersh, Jeff. 1999. "Capitalism Goes Green." Amicus Journal, spring, pp. 37–41.

Gigerenzer, Gerd. 1991. "How to Make Cognitive Illusions Disappear: Beyond 'Heuristics and Biases.'" European Review of Social Psychology 2:83–115.

———. 2007. Gut Feelings: The Intelligence of the Unconscious. New York: Viking.

Gigerenzer, Gerd; Peter Todd; the ABC Group. 1999. Simple Heuristics That Make Us Smart. New York: Oxford University Press.

Gill, Richardson. 2000. The Great Maya Droughts. Albuquerque: University of New Mexico Press.

Gleick, Peter. 2004. The World's Water, 2004–2005. Washington, DC: Island Press.

Goering, Laurie, and Alex Rodriguez. 2008. "Rich Nations Race to Secure Farmland around the Planet." Seattle Times, Dec. 23, A11, from Chicago Tribune.

Goldberg, Jonah. 2006. "That's How You Keep Them Down on the Farm." Seattle Times (syndicated column), Aug. 7, B5.

Goldhagen, Daniel. 1996. Hitler's Willing Executioners. New York: Knopf.

Goldman, Benjamin. 1993. Not Just Prosperity: Achieving Sustainability with Environmental Justice. Washington, DC: National Wildlife Federation.

———. 1996. What Is the Future of Environmental Justice? Oxford: Blackwell.

Goodstein, Eban. 1999. The Trade-Off Myth: Fact and Fiction about Jobs and the Environment. Washington, DC, and Covelo, CA: Island Press.

Goody, Jack. 1993. The Culture of Flowers. Cambridge: Cambridge University Press.

Gore, Al. 2007. The Assault on Reason. New York: The Penguin Press.

Gottlieb, Robert, and Andrew Fisher. 1996. "Community Food Security and Environmental Justice: Searching for a Common Discourse." Agriculture and Human Values 13:23–32.

Goullart, Peter. 1955. Forgotten Kingdom. London: John Murray.

Grain. 2008. "Making a Killing from Hunger." Against the Grain, www.grain.org, retrieved April 30, 2008.

Grenberg, Jeanine M. 1999. "Anthropology from a Metaphysical Point of View." Journal of the History of Philosophy 37:91–115.

Grove, A. T., and Oliver Rackham. 2001. The Nature of the Mediterranean World. New Haven: Yale University Press.

Grunwald, Michael. 2007. "Down on the Farm." Time, Nov. 12, pp. 28–36.

Guha, Ramachandra. 1990. The Unquiet Woods: Economic Change and Peasant Resistance in the Himalaya. Berkeley: University of California Press.

———. 1993. Social Ecology. Delhi: Oxford University Press in India.

Habermas, Jurgen. 1984 (Ger. orig. 1981). The Theory of Communicative Action. Thomas McCarthy (tr.). Boston: Beacon.

Hadot, Pierre. 2002 (Fr. orig. 1995). What Is Ancient Philosophy? Michael Chase (tr.). Cambridge, MA: Harvard University Press.

Haenn, Nora. 2005. Fields of Power, Forests of Discontent: Culture, Conservation, and the State in Mexico. Tucson: University of Arizona Press.

Haidt, Jonathan. 2007. "The New Synthesis in Moral Psychology." Science 36:998–1002.

Hames, Raymond. 2007. "The Ecologically Noble Savage Debate." Annual Review of Anthropology 36:177–190.

Hancock, Graham. 1989/1991. Lords of Poverty. London: Mandarin.

Handwerker, W. Penn. 1997. "Universal Human Rights and the Problem of Unbounded Cultural Meaning." American Anthropologist 99:799–809.

Hardin, Garrett. 1968. "The Tragedy of the Commons." Science 162:1243–1248.

———. 1991. "The Tragedy of the Unmanaged Commons: Population and the Disguises of Providence." In Commons without Tragedy, Robert V. Andelson (ed.). Pp. 162–185. Savage, MD: Barnes and Noble.

Hardin, Lowell S. 2008. "Bellagio 1969: The Green Revolution." Nature 455:470–471.

Harich, Jack. 2007. The Dueling Loops of the Political Powerplace. Clarkston, GA: thwink.

Harremoës, Poul; David Gee; Malcolm MacGarvin; Andy Stirling; Jane Keys; Brian Wynne; Sofia Guedes Vaz (eds.). 2002. The Precautionary Principle in the 20th Century: Late Lessons from Early Warnings. London: Earthscan.

Harris, Edward. 2007. "On the Rise." Associated Press article, as seen in Seattle Times, June 30, p. A3.

He Bochuan. 1991. China on the Edge: The Crisis of Ecology and Development. San Francisco: China Books and Periodicals.

Headland, Thomas. 1997. "Revisionism in Ecological Anthropology." Current Anthropology 38:605–630.

Healey, Jon. 2006. "A Gulf Ghost Town." Los Angeles Times, Feb. 26, pp. M1, M6.

Heather, Peter. 2006. The Fall of the Roman Empire: A New History of Rome and the Barbarians. Oxford: Oxford University Press.

Hedrick, Kimberly. 2007. Our Way of Life: Identity, Landscape, and Conflict. Ph.D. thesis, Dept. of Anthropology, University of California, Riverside.

Heller, Michael, and Rebecca Eisenberg. 1998. "Can Patents Deter Innovation? The Anticommons in Biomedical Research." Science 280:701.

Helvarg, David. 1997. The War against the Greens: The Wise Use Movement, the New Right, and Anti-Environmental Violence. San Francisco: Sierra Club.

Henrich, Joseph; Robert Boyd; Samuel Bowles; Colin Camerer; Ernst Fehr; Herbert Gintis (eds.). 2004. Foundations of Human Sociality: Economic Experiments and Ethnographic Evidence from Fifteen Small-Scale Societies. New York: Oxford University Press.

Herder, Johann Gottfried. 2002. Philosophical Writings. Michael N. Forster (ed./tr.). Cambridge: Cambridge University Press.

Herrnstein, Richard, and Charles Murray. 1994. The Bell Curve. New York: Free Press.

Hess, Karl, Jr. 1992. Visions upon the Land: Man and Nature on the Western Range. Washington, DC, and Covelo, CA: Island Press.

Hinde, Robert. 2007. Bending the Rules: Morality in the Modern World from Relationships to Politics and War. Oxford: Oxford University Press.

Hobbes, Thomas. 1950 [1651]. Leviathan. New York: Dutton.

Hobsbawm, Eric, and Terence Ranger (eds.). 1982. The Invention of Tradition. Cambridge: Cambridge University Press.

Hofrichter, R. 1993. Toxic Struggles: The Theory and Practice of Environmental Justice. Philadelphia: New Society Publishers

Holdren, John. 2008. "Science and Technology for Sustainable Well-Being." Science 319:424–434.

Homer-Dixon, Thomas. 1999. Environment, Scarcity, and Violence. Princeton: Princeton University Press.

Huber, Peter. 1999. Hard Green: Saving the Environment from the Environmentalists—A Conservative Manifesto. New York: Basic Books.

Huber, Toni. 1999. The Cult of Pure Crystal Mountain. New York: Oxford University Press.

Hume, David. 1969 (1739–1740). A Treatise of Human Nature. New York: Penguin.

Humphrey, Caroline, with Urgunge Onon. 1996. Shamans and Elders: Experience, Knowledge, and Power among the Daur Mongols. Oxford: Oxford University Press.

Humphreys, Macartan; Jeffrey D. Sachs; Joseph E. Stiglitz (eds.). 2007. Escaping the Resource Curse. New York: Columbia University Press.

Ibn Khaldun. 1957. The Muqaddimah. Franz Rosenthal (tr./ed.). New York: Pantheon.

Izaak, Robert. 1999. Green Logic: Ecopreneurship, Theory and Ethics. West Hartford, CT: Kumarian Press.

Jaffee, Daniel. 2007. Brewing Justice: Fair Trade Coffee, Sustainability, and Survival. Berkeley: University of California Press.

Janzen, Daniel. 1998. "Gardenification of Wildland Nature and the Human Footprint." Science 279:1312–1313.

Jenkins, Virginia S. 1994. The Lawn: A History of an American Obsession. Washington, DC: Smithsonian Institution Press.

Johnston, Barbara Rose (ed.). 1994. Who Pays the Price? The Sociocultural Context of Enviromental Crisis. Washington, DC: Island Press.

———. 1997. Life and Death Matters: Human Rights and the Environment at the End of the Millennium. Walnut Creek, CA: AltaMira (Sage).

Johnston, David Cay. 2003. Perfectly Legal: The Covert Campaign to Rig Our Tax System to Benefit the Super Rich—and Cheat Everybody Else. New York: Portfolio.

————. 2007. Free Lunch: How the Richest Americans Enrich Themselves at Government Expense (and Stick You with the Bill). New York: Penguin.

Jost, John T. 2006. "The End of the End of Ideology." American Psychologist 61:651–670.

Juhasz, Antonia. 2008. The Tyranny of Oil: The World's Most Powerful Industry—and What We Must Do to Stop It. New York: William Morrow (HarperCollins).

Kant, Immanuel. 2002 (Ger. orig. late 18th century). Groundwork for the Metaphysics of Morals. Allen W. Wood (ed./tr.).. New Haven: Yale University Press.

Kareiva, Peter; Amy Chang; Michelle Marvier. 2008. "Development and Conservation Goals in World Bank Projects." Science 321:1638–1639.

Karl, Terry Lynn. 2007. "Ensuring Fairness: The Case for a Transparent Fiscal Social Contract." In Escaping the Resource Curse. Macartan Humphreys, Jeffrey D. Schas and Joseph E. Stiglitz, (eds.). New York: Columbia University Press. Pp. 256–285.

Kay, Charles, and Randy Simmons (eds.). 2002. Wilderness and Political Ecology. Salt Lake City: University of Utah Press.

Kemper, Theodore D. 2006. "Power and Status and the Power-Status Theory of Emotions." In Handbook of the Sociology of Emotions. Jan E. Stets and Jonathan Turner, (eds.). New York: Springer. Pp. 87–113.

Kenin-Lopsan, Mongush B. 1997. Shamanic Songs and Myths of Tuva. Mihály Hoppál (ed./tr.). Budapest: Akadémiai Kiadó.

Kiers, E. Toby; Roger R. B. Leakey; Anne-Marie Izac; Jack A. Heinemann; Erika Rosenthal; Dev Nathan; Janice Jiggins. 2008. "Agriculture at a Crossroads." Science 320:320–321.

Kirch, Patrick V. 1994. The Wet and the Dry: Irrigation and Agricultural Intensification in Polynesia. Chicago: University of Chicago Press.

————. 1997. "Microcosmic Histories." American Anthropologist 99:31–42.

————. 2007. "Hawaii as a Model System for Human Ecodynamics." American Anthropologist 109:8–26.

Kirk, Ruth. 1986. Wisdom of the Elders. Vancouver: University of British Columbia Press.

Kirsch, Stuart. 2006. Reverse Anthropology. Stanford: Stanford University Press.

Klein, Naomi. 2007. The Shock Doctrine. New York: Metropolitan Books (Henry Holt).

Knoke, Thomas; Bernd Stimm; Michael Weber. 2008. "Tropical Farmers Need Productive Alternatives." (Letter.) Nature 452:934.

Kobori, Iwao, and Michael H. Glantz (eds.). 1998. Central Eurasian Water Crisis: Caspian, Aral, and Dead Seas. Tokyo: United Nations University.

Koerper, Henry, and A. L. Kolls. 1999. "The Silphium Motif Adorning Ancient Libyan Coinage: Marketing a Medicinal Plant." Economic Botany 53:133–143.

Kohen, James L. 1995. Aboriginal Environmental Impacts. Sydney: UNSW Press.

Korsgaard, Christine. 1996. The Sources of Normativity. Cambridge: Cambridge University Press.

Kottak, Conrad, and Alberto C. G. Costa. 1993. "Ecological Awareness, Environmentalist Action, and International Conservation Strategy." Human Organization 52:335–343.

Krech, Shepard, III. 1999. The Ecological Indian: Myth and Reality. New York: W. W. Norton.

Kroeger, Otto, and Janet Thuesen. 1988. Type Talk. New York: Dell.

Kropotkin, Petr. 1955. Mutual Aid: A Factor of Evolution. Boston: Extending Horizon Books.

Krugman, Paul. 2004. The Great Unraveling: Losing Our Way in the New Century. New York: W. W. Norton.

Kuttner, Robert. 2008. The Squandering of America: How the Failure of Our Politics Undermines our Prosperity. New York: Knopf.

Laird, Sarah (ed.). 2002. Biodiversity and Traditional Knowledge: Equitable Partnerships in Practice. London: Earthscan.

Lakoff, George. 2006. Whose Freedom? The Battle over America's Most Important Idea. New York: Farrar, Strauss and Giroux.

Latour, Bruno. 2004. Politics of Nature: How to Bring the Sciences into Democracy. Catherine Porter (tr.). Cambridge, MA: Harvard University Press.

———. 2005. Reassembling the Social: An Introduction to Actor-Network Theory. Oxford: Oxford University Press.

LeDoux, Joseph. 1996. The Emotional Brain: The Mysterious Underpinnings of Emotional Life. New York: Simon & Schuster.

Leff, Enrique (ed.). 2001. Justicia ambiental: construcción y defensa de los nuevos derechos ambientales, culturales y colectivos en América Latina. Mexico City: Centro de Investigaciones Interdisciplinarias en Ciencias y Humanidades.

Lehmann, Scott. 1995. Privatizing Public Lands. New York: Oxford University Press.

Leopold, Aldo. 1949. A Sand County Almanac. New York: Oxford University Press.

Levinas, Emmanuel. 1969 (Fr. orig. 1961). Totality and Infinity. Alphonso Lingis (tr.). Pittsburgh: Duquesne University.

———. 1985 (Fr. orig. 1982). Ethics and Infinity. Richard Cohen (tr.). Pittsburgh: Duquesne University Press.

———. 1989. The Levinas Reader. Sean Hand (ed.). Oxford: Blackwell.

———. 1991 (Fr. orig. 1978). Otherwise than Being or Beyond Essence. Alphonso Lingis (tr.). Dordrecht, Netherlands: Kluwer Academic Publishers.

———. 1994. Outside the Subject. Michael B. Smith (tr.). Stanford: Stanford University Press.

———. 1998a (Fr. orig. 1991). Entre Nous. Michael Smith and Barbara Harshav (tr.). New York: Columbia University Press.

———. 1998b (Fr. orig. 1986). Of God Who Comes to Mind. Bettina Bergo (tr.). Stanford: Stanford University Press.

Lewis, Martin. 1992. Green Delusions. Durham: Duke University Press.

Leys, Simon (Pierre Ryckmans). 1985. The Burning Forest. New York: New Republic Books.

Light, Andrew. 1998. Social Ecology after Bookchin. New York: Guilford.

Lizarralde, Manuel. 1998. "Green Imperialism: Who Will Pay the Cost of Saving the Rainforest?" Paper, American Anthropological Association, Annual Convention, Philadelphia, PA.

Louv, Richard. 2005. Last Child in the Woods: Saving Children from Nature-Deficit Disorder. Chapel Hill: Algonquin Books of Chapel Hill.

Lowe, Celia. 2006. Wild Profusion: Biodiversity Conservation in an Indonesian Archipelago. Princeton: Princeton University Press.

Loyn, David. 2008. "Long Era of Low Food Prices Is Over." BBC News Online, May 29.

Ludwig, Donald; Ray Hilborn; Carl Walters. 1993. "Uncertainty, Resource Exploitation, and Conservation: Lessons from History." Science 260:17, 36.

MacGarvin, Malcolm. 2002. "Fisheries: Taking Stock." In The Precautionary Principle in the 20th Century: Late Lessons from Early Warnings. Poul Harremoës, David Gee, Malcolm MacGarvin, Andy Stirling, Jane Keys, Brian Wynne, Sofia Guedes Vaz (eds.). Pp. 10–25. London: Earthscan.

MacIntyre, Alasdair. 1984. After Virtue: A Study in Moral Theory. Notre Dame, IN: Notre Dame University Press.

———. 1988. Whose Justice? Whose Rationality? Notre Dame, IN: Notre Dame University Press.

Marcus, George E. 2002. The Sentimental Citizen: Emotion in Democratic Politics. University Park, PA: Pennsylvania State University Press.

Markowitz, Gerald, and David Rosner. 2002. Deceit and Denial: The Deadly Politics of Industrial Pollution. Berkeley: University of California Press.

Marks, Robert B. 1997. Tigers, Rice, Silk, and Silt: Environment and Economy in Late Imperial South China. New York: Cambridge University Press.

Marris, Emma. 2008. "Bagged and Boxed: It's a Frog's Life." Nature 452:394–395.

Martinez-Alier, J. 2001. "Mining Conflicts, Environmental Justice, and Valuation." Journal of Hazardous Materials 86:153–170.

Marzluff, John, and Russell Balda. 1992. The Pinyon Jay. London: T. and A. Poyser.

Mauss, Marcel. 1990 (Fr. orig. 1925).The Gift. W. D. Halls (tr.). New York: W. W. Norton.

Maybury-Lewis, David. 1998. "Anthropology and Human Rights." Talk, University of California, Riverside.

McCabe, J. Terrence. 1990. "Turkana Pastoralism: A Case Against the Tragedy of the Commons." Human Ecology 18:81–104.

———. 2003. Sustainability and Livelihood Diversification among the Maasai of Northern Tanzania. Human Organization 62(2):100–111

———. 2004. Cattle Bring Us to Our Enemies: Turkana Ecology, Politics, and Raiding in a Disequilibrium System. Ann Arbor: University of Michigan Press.

McCabe, J. Terrence; Scott Perkin; Claire Schofield. 1992. "Can Conservation and Development Be Coupled among Pastoral People? An Examination of the Maasai of the Ngorongoro Conservation Area, Tanzania." Human Organization 51:353–366.

McCay, Bonnie. 1997. Oyster Wars and the Public Trust. Tucson: University of Arizona Press.

McCay, Bonnie, and James Acheson (eds.). 1987. The Question of the Commons. Tucson: University of Arizona Press.

McCay, Bonnie, and Svein Jentoft. 1996. "From the Bottom Up: Participatory Issues in Fisheries Management." Society and Natural Resources 9:237–250.

McEvoy, Arthur. 1986. The Fisherman's Problem. Berkeley: University of California Press.

McGinn, Anne Platt. 1998. Rocking the Boat: Conserving Fisheries and Protecting Jobs. Washington, DC: Worldwatch Institute.

Mead, George Herbert. 1964. The Social Psychology of George Herbert Mead. Anselm Strauss (ed.). Chicago: University of Chicago Press.

Melville, Elinor G. K. 1997. A Plague of Sheep: Environmental Consequences of the Conquest of Mexico. New York: Cambridge University Press.

Mencius. 1970. Mencius. D. C. Lau (tr.). New York: Penguin.

Menzies, Charles R. (ed.). 2006. Traditional Ecological Knowledge and Natural Resource Management. Lincoln: University of Nebraska Press.

Merchant, Carolyn. 1992. Radical Ecology. London: Routledge.

Merleau-Ponty, Maurice. 2003 (Fr. orig. 1995). Nature. Robert Vallier (tr.). Evanston, IL: Northwestern University Press.

Metzo, Katherine R. 2005. "Articulating a Baikal Environmental Ethic." Anthropology and Humanism 30:39–54.

Michaels, David. 2008. Doubt Is Their Product: How Industry's Assault on Science Threatens Your Health. New York: Oxford University Press.

Micklin, Philip, and Nikolay V. Aladin. 2008. "Reclaiming the Aral Sea." Scientific American, April, 64–71.

Midgeley, Mary. 1995. "Duties Concerning Islands." In Environmental Ethics, Robert Elliott (ed.). Oxford: Oxford University Press. Pp. 89–103.

Millennium Ecosystem Assessment. 2005. MEA Synthesis Report. www.maweb.org.

Miller, Christopher. 1999. Environmental Rights: Critical Perspectives. London: Routledge.

Miller, T. Christian. 1998a. "Developer Money Pours into Campaign Coffers of Officials." Los Angeles Times, Dec. 27, p. 27.

———. 1998b. "A Growth Plan Run Amok." Los Angeles Times, Dec. 27, pp. 1, 27–29.

Mills, C. Wright. 1959. The Sociological Imagination. New York: Oxford University Press.

Mintz, Sidney. 1985. Sweetness and Power. New York: Viking Penguin.

Mokhiber, Russell, and Robert Weissman. 1999a. Corporate Predators: The Hunt for Megaprofits and the Attack on Democracy. Monroe, MN: Common Courage Press.

———. 1999b. Wealth Distribution in an 11,000 Dow. Syndicated column, distributed electronically.

Moll, Jorge, and Ricardo de Oliveira-Souza. 2008. "When Morality Is Hard to Like." Scientific American Mind, Feb.-Mar., 30–35.

Mosse, David. 2006. "Rule and Representation: Transformations in the Governance of the Water Commons in British South India." Journal of Asian Studies 65:61–90.

Murphy, Earl. 1967. Governing Nature. Chicago: Aldine.

Myerhoff, Barbara. 1978. Number Our Days. New York: Simon and Schuster.

Myers, Isabel Briggs, with Peter B. Myers. 1980. Gifts Differing. Palo Alto, CA: Consulting Psychologists Press.

Myers, Norman. 1997. "Consumption: Challenge to Sustainable Development...." Science 276:53–57.

———. 1999. "What We Must Do to Counter the Biotic Holocaust." International Wildlife, March-April, 30–39.

———. 2008. "Perverse Priorities." World Conservation (IUCN), May, pp. 6–7.

Myers, Norman, with Jennifer Kent. 1998. Perverse Subsidies: Tax Dollars Undercutting Our Economics and Environments Alike. Winnipeg: International Institute for Sustainable Development.

Nabhan, Gary. 1987. Gathering the Desert. Tucson: University of Arizona Press.

———. 1997. Cultures of Habitat: On Nature, Culture and Story. Washington, DC: Counterpoint.

Nadasdy, Paul. 2004. Hunters and Bureaucrats: Power, Knowledge, and Aboriginal-State Relations in the Southwest Yukon. Vancouver: University of British Columbia Press.

———. 2007. "The Gift of the Animals: The Ontology of Hunting and Human-Animal Sociality." American Ethnologist 34:25–47.

Nadeau, Robert. 2008. "The Economist Has No Clothes." Scientific American, April, 42.

Netting, Robert. 1981. Balancing on an Alp. Cambridge: Cambridge University Press.

———. 1993. Smallholders, Householders. Stanford: Stanford University Press.

Newcomb, Steven. 1994. "Traditional Native Law: Concepts for an Environmentally Sound Way of Life." Eugene, OR: Indigenous Law Institute.

Normile, Dennis. 2008a. "As Food Prices Rise, U. S. Support for Agricultural Centers Wilts." Science 320:303.

———. 2008b. "Reinventing Rice to Feed the World." Science 321:330–333.

North, Douglass. 1990. Institutions, Institutional Change, and Economic Performance. Cambridge: Cambridge University Press.

Norton, Bryan. 1991. Towards Unity among Environmentalists. New York: Oxford University Press.

Nyerges, Endre (ed.). 1997. The Ecology of Practice. New York: Gordon and Breach.

Oliveira, Paulo J. C.; Gregory P. Asner; David E. Knapp; Angélica Almeyda; Ricardo Galván-Gildemeister; Sam Keene; Rebecca F. Raybin; Richard C. Smith. 2007. "Land-Use Allocation Protects the Peruvian Amazon." Science 317:1233–1236.

Olson, Mancur. 1965. The Logic of Collective Action. Cambridge, MA: Harvard University Press.

Orians, Gordon, and Judith Heerwagen. 1992. "Evolved Responses to Landscapes." In The Adapted Mind, Jerome Barkow, Leda Cosmides, John Tooby (eds.). Pp. 555–580. New York: Oxford University Press.

Ostrom, Elinor. 1990. Governing the Commons: The Evolution of Institutions for Collective Action. New York: Cambridge University Press.

———. 1999. Executive Summary, Self-Governance and Forest Resources. CIFOR Occasional Paper No. 20.

Ostrom, Elinor; Marco Janssen; John M. Anderies. 2007. "Going Beyond Panaceas." Proceedings of the National Academy of Sciences 104:15176–15178.

Painter, Michael, and William Durham (eds.). 1995. The Social Causes of Environmental Destruction in Latin America. Ann Arbor: University of Michigan.

Palaniappan, Meena; Emily Lee; Andrea Samulon. 2006. "Environmental Justice and Water." In The World's Water, 2006–2007: The Biennial Report on Freshwater Resources, Peter H. Gleick (ed.). Pp. 117–144. Washington: Island Press.

Pastor, Manuel. 2001. Building Social Capital to Protect Natural Capital: The Quest for Environmental Justice. Amherst, MA: University of Massachusetts, Political Economy Research Institute, working paper 11.

Pastor, Manuel; James L. Sadd; Rachel Morello-Frosch. 2002. "The Continuing Environmental Justice Debate. Who's Minding the Kids? Pollucion [sic], Public Schools, and Environmental Justice in Los Angeles." Social Science Quarterly 83:263–280.

Peet, Roger, and Michael Watts (eds.). 1996. Liberation Ecologies: Environment, Development, Social Movements. London: Routledge.

Pellow, D. 2004. Garbage Wars: The Struggle for Environmental Justice in Chicago. Cambridge, MA: MIT Press.

Pellow, David Naguid, and Robert J. Brulle (eds.). 2005. Power, Justice, and the Environment: A Critical Appraisal of the Environmental Justice Movement. Cambridge, MA: MIT Press.

Pellow, David Naguid, and Lisa Sun-He Park. 2002. The Silicon Valley of Dreams: Environmental Injustice, Immigrant Workers, and the High-Tech Global Economy. New York: New York University Press.

Peluso, Nancy Lee. 1992. Rich Forests, Poor People: Resource Control and Resistance in Java. Berkeley: University of California Press.

Perkins, John. 2004. Confessions of an Economic Hit Man. New York: Penguin.

Pervin, Lawrence (ed.). 1990. Handbook of Personality: Theory and Research. New York: Guilford.

Petrinovitch, Lewis. 1995. Human Evolution, Reproduction, and Morality. New York: Plenum.

Phillips, Kevin. 2008. Bad Money: Reckless Finance, Failed Politics and the Global Crisis of American Capitalism. New York: Viking.

Picolotti, Romina, and Jorge Daniel Taillant (eds.). 2003. Linking Human Rights and the Environment. Tucson: University of Arizona Press.

Pierce, David W., and R. Kerry Turner. 1990. Economics of Natural Resources and the Environment. Baltimore: Johns Hopkins University Press.

Pierotti, Raymond, and Daniel R. Wildcat. 1999. "Traditional Knowledge, Culturally-based World-views and Western Science." In Cultural and Spiritual Values of Biodiversity, Darrell Posey (ed.). Pp. 192–199. London and Nairobi: UN Environment Programme.

Pimentel, David. 1999. "Human Resource Use, Population Growth, and Environmental Destruction." Bulletin of the Ecological Society of America 80:88–91.

Pinkerton, Evelyn (ed.). 1989. Cooperative Management of Local Fisheries: New Directions for Improved Management and Community Development. Vancouver: University of British Columbia Press.

Pinkerton, Evelyn, and Martin Weinstein. 1995. Fisheries that Work. Vancouver, BC: David Suzuki Foundation.

Pollan, Michael. 1999. "In a Town Where Lawns Are Banned, Wildlife and Community Pride Flourish." National Wildlife 37:4:14–16.

———. 2006. The Omnivore's Dilemma: A Natural History of Four Meals. New York: Penguin.

Pollini, Jacques. 2007. Slash-and-Burn Cultivation and Deforestation in the Malagasy Rain Forests: Representations and Realities. Ph.D. dissertation, Cornell University, Program in Natural Resources.

Pombo, Richard, and Joseph Farah. 1996. This Land Is Our Land: How to End the War on Private Property. New York: St. Martin's.

Ponting, Clive. 1991. A Green History of the World. New York: Penguin.

Pooley, Eric. 2007. "The Fox in the Henhouse." Time, June 30, pp. 32–38.

Posey, Darrell Addison. 2004. Indigenous Knowledge and Ethics: A Darrell Posey Reader. New York: Routledge.

Posey, Darrell (ed.). 1999. Cultural and Spiritual Values of Biodiversity. London and Nairobi: UN Environment Programme.

Price, Michael. 2008. "Making Sense of Dollars and Cents." Monitor (APA), Feb., 34–36.

Primack, Richard; David Bray; Hugo A. Galletti; Ismael Ponciano (eds.). 1998. Timber, Tourists and Temples: Conservation and Development in the Maya Forest of Belize, Guatemala, and Mexico. Washington, DC: Island Press.

Putnam, Robert; Robert Leonardi; Raffaella Nanetti. 1992. Making Democracy Work: Civic Traditions in Modern Italy. Princeton: Princeton University Press.

Putnam, Robert. 2001. Bowling Alone: The Collapse and Revival of American Community. New York: Simon and Schuster.

Pye-Smith, Charlie. 2002. The Subsidy Scandal. London: Earthscan.

Pyne, Stephen. 1991. Burning Bush: A Fire History of Australia. New York: Henry Holt.

Qing, Dai. 1998. The River Dragon Has Come: The Three Gorges Dam and the Fate of China's Yangtze River and Its Peoples. Armonk, NY: M. E. Sharpe.

Raganathan, Jai; R. J. Ranjit Daniels; Subash Chandran; Paul Ehrlich; Gretchen Daily. 2008. Sustaining Biodiversity in Ancient Tropical Countryside. Proceedings of the National Academy of Sciences 105:17852–17854.

Ratchnevsky, Paul. 1991. Genghis Khan: His Life and Legacy. Oxford: Basil Blackwell.

Rangan, Haripraya. 1996. "From Chipko to Uttaranchal." Peet and Watts 1996:205–226.

Rappaport, Roy. 1984. Pigs for the Ancestors. 2nd edn. New Haven: Yale University Press.

Rawls, John. 1971. A Theory of Justice. Cambridge, MA: Harvard University Press.

———. 2005. A Theory of Justice. New edn. Cambridge, MA: Belknap Press.

Rea, Amadeo. 1983. Once a River: Bird Life and Habitat Changes on the Middle Gila. Tucson: University of Arizona Press.

Reichhardt, Tony; Erika Check; Emma Marris. 2005. "After the Flood." Nature 437:174–176.

Reinhardt, Forest L. 2000. Down to Earth: Applying Business Principles to Environmental Management. Cambridge, MA: Harvard Business School Press.

Repetto, Robert. 1990. "Deforestation in the Tropics." Scientific American 262, April 1990, 36–42.

———. 1992. "Accounting for Environmental Assets." Scientific American, June, pp. 94–100.

Rich, Bruce. 1994. Mortgaging the Earth: The World Bank, Environmental Impoverishment, and the Crisis of Development. Boston: Beacon.

Rich, Frank. 2006. The Greatest Story Ever Sold: The Decline and Fall of Truth from 9/11 to Katrina. New York: Penguin.

Ridley, Matt. 1996. The Origins of Virtue. New York: Penguin.

Riesman, David; Reuel Denny; Nahum Glazer. 1953. The Lonely Crowd: A Study of the Changing American Character. New Haven: Yale University Press.

Rifkin, Jeremy. 1992. Beyond Beef. New York: Dutton.

Ripe, Cherry. 1994. Dying of Starvation at the Supermarket. Ms. Roberts, J. Timmons, and Melissa M. Toffolon-Weiss. 2001. Chronicles from the Environmental Justice Frontline. New York: Cambridge University Press.

Robbins, Paul. 2004. Political Ecology. Oxford: Blackwell.

Rodman, David Malin. 1999. "Building a Sustainable Society." In: State of the World 1999. Lester Brown, Christopher Flavin, and Hilary French (eds.). New York: W. W. Norton.

Rolston, Holmes, III. 1988. Environmental Ethics: Duties to and Values in the Natural World. Philadelphia: Temple University Press.

Rose, Deborah Bird. 2000. Dingo Makes Us Human: Life and Land in an Australian Aboriginal Culture. New York: Cambridge University Press.

Rose, Deborah Bird. 2005. Reports from a Wild Country: Ethics for Decolonisation. Sydney: University of New South Wales Press.

Rosenberg, A. A.; M. J. Fogarty; M. P. Sissenwine; J. R. Beddington; J. G. Shepherd. 1993. "Achieving Sustainable Use of Renewable Resources." Science 262:828–829.

Rosenberg, Alexander. 1992. Economics—Mathematical Politics or Science of Diminishing Returns: Chicago: University of Chicago Press.

Rosenthal, Franz (tr./ed.) 1958. Ibn Khaldun, the Muqaddimah: An Introduction to History. New York: Bollingen.

Rossi, Melisa. 2006. What Every American Should Know about Who's Really Running the World. New York: Plume (Penguin Group).

Ross-Ibarra, Jeffrey, and Alvaro Molina-Cruz. 2002. "The Ethnobotany of Chaya (*Cnidoscolus aconitifolius* ssp. *aconitifolius* Breckon): A Nutritious Maya Vegetable." Economic Botany 56:350–365.

Rothkopf, David. 2008. Superclass: The Global Elite and the World They Are Making. New York: Farrar, Straus and Giroux.

Roush, Wade. 1997. "Putting a Price Tag on Nature's Bounty." Science 276:1029.

Rousseau, J.-J. 1983. On the Social Contract: Discourse on the Origin of Inequality (and) Discourse on Political Economy. Donald Cress (tr.). Indianapolis: Hackett.

Roux, Jean. 1966. Faune et flore sacrées dans les sociétés altaïques. Paris: A. Maisonneuve.

Rummel, Rudolph. 1998. Statistics of Democide. Munster: Lit Verlag.

Sachs, Aaron. 1995. Eco-Justice: Linking Human Rights and the Environment. Washington: World Watch Institute.

Safina, Carl. 1997. Song for the Blue Ocean. New York: Henry Holt and Co.

Salaman, Redcliffe. 1985. The History and Social Influence of the Potato. 2nd edn, with introduction by J. Hawkes. Cambridge: Cambridge University Press.

Salazar, D. J. "Environmental Justice and a People's Forestry." Journal of Forestry 94:32–38.

Samardzija, Michael R. 2007. "The Obvious War." Science 315:190–191.

Sandler, Ronald, and Philip Cafaro (eds.). 2005. Environmental Virtue Ethics. Oxford: Rowman and Littlefield.

Satterfield, Terre A.; C. K. Mertz; Paul Slovic. 2004. "Discrimination, Vulnerability, and Justice in the Face of Risk." Risk Analysis 24:115–129.

Satterfield, Theresa A. 2000. "Risk, Remediation and the Stigma of a Technological Accident in an African-American Community." Human Ecology Review 7:1–11.

Sayre, Nathan F. 2002. Ranching, Endangered Species, and Urbanization in the Southwest: Species of Capital. Tucson: University of Arizona Press.

Scanlon, Tom. 1998. What We Owe to Each Other. Cambridge, MA: Harvard University Press.

Schama, Simon. 1995. Landscape and Memory. New York: Alfred A. Knopf.

Schelhas, John. 2002. "Race, Ethnicity, and Natural Resources in the United States: A Review." University of New Mexico, School of Law, Natural Resources Journal, 42:723–763.

Schelhas, John, and Max Pfeffer. 2008. Saving Forests, Protecting People? Environmental Conservation in Central America. Lanham, MD: AltaMira, a Division of Rowman & Littlefield Publishers.

Schlosberg, D. 1999. Environmental Justice and the New Pluralism. New York: Oxford University Press.

Schmidtz, David. 1991. The Limits of Government: An Essay on the Public Goods Argument. Boulder, CO: Westview.

Schultz, Theodore. 1968. Agriculture and Economic Growth. New York: McGraw-Hill.

Science. 2007. "Keeping Tabs on Killer Tabbies." Science 315:167.

Scott, James. 1985. Weapons of the Weak. New Haven: Yale University Press.

———. 1990. Domination and the Arts of Resistance. New Haven: Yale University Press.

———. 1998. Seeing like a State. New Haven: Yale University Press.

Scriven, Tal. 1997. Wrongness, Wisdom, and Wilderness. Albany: SUNY Press.

Scudder, Thayer. 2005. The Future of Large Dams: Dealing with Social, Environmental, Institutional and Political Costs. London: Earthscan.

Sen, Amartya. 1973. On Economic Inequality. Oxford: Oxford University Press.

———. 1984. Resources, Values, and Development. Cambridge, MA: Harvard University Press.

———. 1992. Inequality Reconsidered. Cambridge: Harvard University Press and Russell Sage Foundation.

———. 1993. "The Economics of Life and Death." Scientific American, May 1993, pp. 40–47

———. 1997. Hunger in the Contemporary World. London: London School of Economics.

———. 1999. Commodities and Capabilities. Delhi: Oxford University Press.

———. 2001. Development as Freedom. New York: Oxford University Press.

Shandra, John M.; Eran Shor; Gary Maynard; Bruce London. 2008. "Debt, Structural Adjustment, and Deforestation: A Cross-National Study." Journal of World-Systems Research 14:1–20.

Shell, Ellen Ruppel. 2008. "Trashed." Audubon Magazine, May-June, 90–96.

Sheridan, Thomas. 1988. Where the Dove Calls. Tucson: University of Arizona Press.

Shiva, Vandana. 1997. Biopiracy: The Plunder of Nature and Knowledge. Boston: South End Press.

Shrader-Frechette, K. S. 1991. Risk and Rationality: Philosophical Foundations for Populist Reforms. Berkeley: University of California Press.

———. 2002. Environmental Justice: Creating Equality, Reclaiming Democracy. New York: Oxford University Press.

Sidgwick, Henry. 1902. Outlines of the History of Ethics. London: MacMillan.

———. 1981 [1907]. The Methods of Ethics. New York: Hackett.

Sierra [magazine]. 2006. "As the World Warms." Sierra, Feb., p. 13.

Sirota, David. 2007. Hostile Takeover: How Big Money and Corruption Conquered Our Government—and How We Take It Back. New York: Three Rivers Press.

———. 2008. "A Different Kind of Democracy." Syndicated column (retrieved in *Seattle Times*, May 26, p. B5).

Smil, Vaclav. 1984. The Bad Earth. Armonk, NY: M. E. Sharpe.

———. 1993. China's Environmental Crisis. Armonk, NY: M. E. Sharpe.

———. 2004. China's Past, China's Future: Energy, Food, Environment. London: RoutledgeCurzon.

Smith, Adam. 1910 (orig. 1776). The Wealth of Nations. London: J. M. Dent.

Smith, Eric Alden, and Bruce Winterhalder (eds.). 1992. Evolutionary Ecology and Human Behavior. New York: Aldine de Gruyter.

Smith, Eric Alden, and M. Wishnie. 2000. "Conservation and Subsistence in Small-Scale Societies." Annual Review of Anthropology 25:105–126.

Snodgrass, Jeffrey, and Kristina Tiedje. 2008. "Introduction: Indigenous Nature Reverence and Conservation—Seven Ways of Transcending an Unnecessary Dichotomy." Journal for the Study of Religion, Nature and Culture 2:6–29.

Southgate, Charles. 1998. Tropical Forest Conservation: An Economic Assessment of the Alternatives in Latin America. New York: Oxford University Press.

Sponsel, Leslie E. 2001a. "Do Anthropologists Need Religion, and Vice Versa? Adventures and Dangers in Spiritual Ecology." *In* New Directions in Anthropology and Environment: Intersections. C. Crumley (ed.). Pp. 177–200. New York: AltaMira.

———. 2001b. "Is Indigenous Spiritual Ecology Just a New Fad? Reflection on the Historical and Spiritual Ecology of Hawai'i." *In* Indigenous Traditions and Ecology: The Interbeing of Cosmology and Community, John Grim (ed.). Pp. 159–174. Cambridge, MA: Harvard University Press.

Srinivasan, U. Thara; Susan P. Carey; Eric Hallstein; Paul A. T. Higgins; Amber C. Kerr; Laura E. Koteen; Adam B. Smith; Reg Watson; John Harte; Richard B. Norgaard. 2008. "The Debt of Nations and the Distribution of Ecological Impacts from Human Activities." Proceedings of the National Academy of Sciences 105:1768–1773.

Stauber, John, and Sheldon Rampton. 1995. Toxic Sludge is Good for You: Lies, Damn Lies and the Public Relations Industry. Monroe, Maine: Common Courage.

Stedman, John Gabriel. 1988. Narrative of a Five Years Expedition against the Revolted Negroes of Surinam. Transcribed for the First Time from the Original 1790 Manuscript. Richard and Sally Price (eds./intr.). Baltimore: Johns Hopkins University Press.

Steinberg, Ted. 2006. Acts of God: The Unnatural History of Natural Disaster in America. New York: Oxford University Press.

Stern, Theodore. 1965. The Klamath Tribe: A People and Their Reservation. Seattle: University of Washington Press.

Stevens, Wallace. 1952. The Man with the Blue Guitar. New York: Alfred A. Knopf.

Stiassny, Melanie, and Axel Meyer. 1999. "Cichlids of the Rift Lakes." Scientific American, Feb., 64–69.

Stiglitz, Joseph E. 2003. Globalization and Its Discontents. New York: W. W. Norton.

Stix, Gary. 2006. "Owning the Stuff of Life." Scientific American, Feb., 76–83.

Stoll, David. 1993. Between Two Armies in the Ixil Towns of Guatemala. New York: Columbia University Press.

———. 1999. Rigoberta Menchu and the Story of All Poor Guatemalans. Boulder, CO: Westview.

Stone, Richard. 2007. "Subduing Poachers, Ducking Insurgents to Save a Splendid Bird." Science 317:592–593.

Stonich, Susan C. 1993. "I Am Destroying the Land!" The Political Ecology of Poverty and Environmental Destruction in Honduras. Boulder: Westview, 1993.

Strang, Veronica. 1997. Uncommon Ground. Oxford: Berg.

Suliman, Mohamed (ed.). 1999. Ecology, Politics, and Violent Conflict. London: Zed Books.

Sumaila, U. Rashid, and Daniel Pauly. 2007. "All Fishing Nations Must Unite to Cut Subsidies." Letter. Nature 450:945.

Szasz, Andrew. 1994. Ecopopulism: Toxic Waste and the Movement for Environmental Justice. Minneapolis: University of Minnesota Press.

Tarrant, Michael A., and H. Ken Cordell. 1999. "Environmental Justice and Spatial Distribution of Outdoor Recreation Sites: An Application of Geographic Information Systems." Journal of Leisure Research 31.

Taylor, Michael. 2006. Rationality and the Ideology of Disconnection. New York: Cambridge University Press.

Taylor, Shelley E. 1989. Positive Illusions: Creative Self-Deception and the Healthy Mind. New York: Basic Books.

Terborgh, John. 1989. Where Have All the Birds Gone? Princeton: Princeton University Press.

———. 1999. Requiem for Nature. Washington, DC: Island Press.

Thompson, E. P. 1963. The Making of the English Working Class. New York: Random House.

Thu, Kendall. 1998. "The Health Consequences of Industrialized Agriculture for Farmers in the United States. Human Organization 57:335–341.

Tisdell, Clem. 1999. Biodiversity, Conservation and Sustainable Development. Cheltenham, UK: Edward Elgar.

Totman, Conrad. 1989. The Green Archipelago: Forestry in Preindustrial Japan. Berkeley: University of California Press.

———. 1995. The Lumber Industry in Early Modern Japan. Honolulu: University of Hawai'i Press.

Totten, Arthur, and Bill Dickerson (eds.). 1998. Final Guidance for Incorporating Environmental Justice Concerns in EPA's NEPA Compliance Analyses. Washington, DC: United States Government, Environmental Protection Agency.

Travis, John. 2005. "Scientists' Fears Come True as Hurricane Floods New Orleans." Science 309:1656–1659.

Tuchman, Barbara. 1984. The March of Folly: From Troy to Vietnam. New York: Knopf.

Tucker, Mary Evelyn, and John A. Grim (eds.). 1994. Worldviews and Ecology: Religion, Philosophy, and the Environment. Maryknoll, NY: Orbis Books.

Tulalip Tribes of Washington. 2007. Huchoosedah Protection Act. Ms.

Turner, Nancy J. 2005. The Earth's Blanket. Vancouver: Douglas and MacIntyre; Seattle: University of Washington Press.

Turner, Robin Lanette, and Diana Pei Wu. 2002. Environmental Justice and Environmental Racism: An Annotated Bibliography and General Overview, Focusing on U. S. Literature, 1996–2002. Berkeley: Institute of International Studies, University of California.

Uhl, Christopher. 1998. "Conservation Biology in Your Own Front Yard." Conservation Biology 12:1175–1177.

United States Government, Environmental Protection Agency. 1994. "Federal Actions to Address Environmental Justice in Minority Populations and Low-Income Populations: Executive Order 12898." Washington, DC: United States Environmental Protection Agency.

Vallette, Jim. 1999. "The Tragic Rise of the New Treasury Secretary." Email posting.

Van Dieren, Wouter (ed.). 1995. Taking Nature into Account: A Report to the Club of Rome. New York: Springer-Verlag.

Van Vugt, Mark; Robert Hogan; Robert B. Kaiser. 2008. "Leadership, Followership, and Evolution." American Psychologist 63:182–196.

Vayda, Andrew, and Bradley B. Walters. 1999. "Against Political Ecology." Human Ecology 27:167–179.

Vincent, Jeffrey, and Theodore Panayatou. 1997. "… Or Distraction?" Science 276:53–57.

Vogel, Joseph Henry. 1994. Genes for Sale: Privatization as a Conservation Policy. New York: Oxford University Press.

———. 2000. The Biodiversity Cartel: Transformation of Traditional Knowledge into Trade Secrets. (CD.) Quito, Ecuador: CARE.

Wagner, Marsden. 2006. Born in the USA: How a Broken Maternity System Must Be Fixed to Put Women and Children First. Berkeley: University of California Press.

Wallerstein, Immanuel. 1976. The Modern World-System: Capitalist Agriculture and the Origins of the European World-Economy in the Sixteenth Century. New York: Academic Press.

Walker, Brian, and David Salt. 2006. Resilience Thinking: Sustaining Ecosystems and People in a Changing World. Washington, DC: Island Press.

Walsh, Mary Williams. 2000. "Boom Time a Bad Time for Poorest, Study Finds." Los Angeles Times, Jan. 19, pp. 1, 12.

Warren, Kay (ed.). 1993. The Violence Within: Cultural and Political Opposition in Divided Nations. Boulder, CO: Westview.

Watson, James L. (ed.). 1998. Golden Arches East. Stanford: Stanford University Press.

Weber, Max. 1946. *From* Max Weber. Hans Gerth and C. Wright Mills (tr./ed.). New York: Oxford University Press.

———. 2002. The Protestant Ethic and the Spirit of Capitalism. Peter Baehr, Gordon Wills (tr.). New York: Penguin.

Webster, David. 2002. The Fall of the Ancient Maya. New York: Thames and Hudson.

Wensveen, Louke van. 2000. Dirty Virtues: The Emergence of Ecological Virtue Ethics. Amherst, NY: Humanity Books.

West, Paige. 2006. Conservation Is Our Government Now: The Politics of Ecology in Papua New Guinea. Durham: Duke University Press.

West, Paige; James Igoe; Dan Brockington. 2006. "Parks and Peoples: The Social Impact of Protected Areas." Annual Review of Anthropology 35:251–277.

Westen, Drew. 2007. The Political Brain: The Role of Emotion in Deciding the Fate of the Nation. New York: PublicAffairs.

White, Leslie. 1949. The Science of Culture. New York: Grove Press.

White, Richard. 1996. "Are You an Environmentalist, or Do You Work for a Living?" *In* Uncommon Ground, William Cronon (ed.). Pp. 171–185. New York: W. W. Norton.

Wilke, Philip J. 1988. "Bow Staves Harvested from Juniper Trees by Indians of Nevada." Journal of California and Great Basin Anthropology 10:3–31.

Wilkinson, Charles. 1992. Crossing the Next Meridian: Land, Water, and the Future of the West. Washington, DC, and Covelo, CA: Island Press.

Williams, Dee Mack. 1996a. "The Barbed Walls of China: A Contemporary Grassland Drama." Journal of Asian Studies 55:665–691.

———. 1996b. "Grassland Enclosure: Catalyst of Land Degradation in Inner Mongolia." Human Organization 55:307–313.

———. 2000. "Representations of Nature on the Mongolian Steppe: An Investigation of Scientific Knowledge Construction." American Anthropologist 102:503–519.

Williams, Roger. 1956. Biochemical Individuality. New York: Wiley.

Wilson, E. O. 1984. Biophilia. Cambridge, MA: Harvard University Press.

Wohl, Ellen. 2005. "Compromised Rivers: Understanding Human Impacts on Rivers in the Context of Restoration." Ecology and Society 10, no. 2, article 2.

Wolf, Eric. 1982. Europe and the "People without History." Berkeley: University of California Press.

Wolfowitz, Paul. 2005. "Free Trade for a Better Future." Los Angeles Times, Dec. 13, B13.

Woodard, Colin. 2007. "In Greenland, an Interfaith Rally for Climate Change." Christian Science Monitor, Sept. 12, retrieved from Yahoo! News Online.

Worm, Boris; Edward B. Barbier; Nicola Beaumont; J. Emmett Duffy; Carl Folke; Benjamin S. Halpern; Jeremy B. C. Jackson; Heike K. Lotze; Fiorenza Micheli; Stephen R. Palumbi; Enric Sala; Kimberley A. Selkoe; John J. Stachowicz; Reg Watson. 2006. "Impacts of Biodiversity Loss on Ocean Ecosystem Services." Science 314:787–790.

Wright, Angus. 1990. The Death of Ramon Gonzalez. Berkeley: University of California Press.

———. 2005. The Death of Ramon Gonzalez. 2nd edn. Berkeley: University of California Press.

Zajonc, Robert. 1980. "Feeling and Thinking: Preferences Need No Inferences." American Psychologist 35:151–175.

Zepezauer, Mark, and Arthur Naiman. 1996. Take the Rich Off Welfare. Cambridge, MA: South End Press.

Zerner, C. (ed.). 2000. People, Plants and Justice: The Politics of Nature Conservation. New York: Columbia University Press.

Zimmerer, Karl S., and Thomas J. Bassett. 2003. Political Ecology: An Integrative Approach to Geography and Environment-Development Studies. New York: Guilford.

Index

About the Author

E. N. ANDERSON is Professor Emeritus of Anthropology at the University of California, Riverside. He has done field work in Hong Kong, Malaysia, Mexico, British Columbia, California, and (for shorter periods) several other areas. He is primarily interested in cultural and political ecology, ethnobiology, and the development of ideas and representations of the environment (especially plants and animals). He has also done research in medical and nutritional anthropology, branching out from a human-ecology focus. He has written several books, including *The Food of China* (Yale University Press, 1988), *Ecologies of the Heart* (Oxford University Press, 1996), *Everyone Eats* (New York University Press, 2005), and *Floating World Lost*, an ethnography of a Hong Kong fishing community (University Press of the South, 2007). He is currently working on the relationship between ideology, resource management, and cultural representations of the environment.